ecture Notes in Mathematics

1481

D. Ferus U. Pinkall
U. Simon B. Wegner (Eds.)

Global Differential Geometry and Global Analysis

Proceedings of a Conference held in
Berlin, 15-20 June, 1990

Springer-Verlag
Berlin Heidelberg New York
London Paris Tokyo
Hong Kong Barcelona
Budapest

Editors

Dirk Ferus
Ulrich Pinkall
Udo Simon
Berd Wegner
Fachbereich Mathematik
Technische Universität Berlin
W-1000 Berlin 12, FRG

Mathematics Subject Classification (1991): 53-06, 58-06, 53A04, 53A07, 53A10, 53A15, 53A60, 53C30, 53C42, 58A50, 58D05, 58E05, 58G25, 58G30

ISBN 3-540-54728-2 Springer-Verlag Berlin Heidelberg New York
ISBN 0-387-54728-2 Springer-Verlag New York Berlin Heidelberg

Printed in Germany

Typesetting: Camera ready by author
Printing and binding: Druckhaus Beltz, Hemsbach/Bergstr.
46/3140-543210 - Printed on acid-free paper

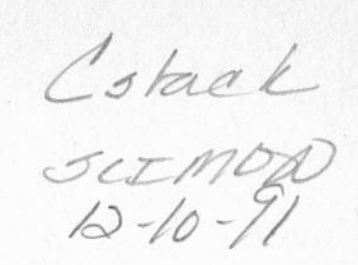

Introduction

This conference continued a long tradition of similar meetings held at TU Berlin in earlier years. It was, however, for two reasons very distinct from its predecessors:

For the first time the EC-Project "Global Analysis, Geometry and Applications" provided the framework, bringing together more than fifty representatives of the twelve participating institutes from all over Western Europe. The exchange between our project and the recent developements in geometry and analysis outside the realms of the EC program was a major objective during the meeting.

The second novelty of this conference was caused by the unexpected political changes in Berlin, Germany and East Europe. Encounters with mathematicians from eastern countries have always been a special feature of the Berlin conferences, but this time there came more than sixty mathematicians from the former DDR and other socialist countries. Many of us had know each other for a long time through our publications, but had never met personally. It became quite obvious, that the conference laid the foundation for a number of future east-west cooperations.

The total number of more than 200 participants was twice that of earlier years. About 40 colleagues came from overseas.

The notes presented in this volume give only an incomplete selection of the topics covered during the conference. We therefore include a list of the participants and of all invited titles.

We are grateful for the support that made this conference possible. Our thanks go to

European Community (Contract SC 1-0039-C AM)
Deutsche Forschungsgemeinschaft
Deutscher Akademischer Austauschdienst
Senat von Berlin
Technische Universität Berlin.

We also thank Springer Verlag for the publication of a third proceedings volume on a Global Geometry and Analysis Conference at the TU Berlin.

Bernd Wegner, Dirk Ferus, Ulrich Pinkall, Udo Simon

Contents

On the minimal hypersurfaces of a locally symmetric manifold

Stefana Hineva, Evgeni Belchev

1. Introduction.

Let M^n be an n-dimensional hypersurface which is minimally immersed in an n+1-dimensional locally symmetric manifold N^{n+1}. Let h be the second fundamental form of this immersion. We denote by S the square of the length of h. In this paper we give conditions in which S is constant and show when S vanishes, i.e. when M^n is totally geodesic.

2. Local formulas for a minimal hypersurface.

In this section we shall compute the Laplacian of the second fundamental form of a minimal hypersurface M^n of a locally symmetric space N^{n+1}. We shall follow closely the exposition in [1], even we shall take some formulas directly from [1]. Let $e_1, e_2, \dots, e_n, e_{n+1}$ be a local frame of orthonormal vector fields in N^{n+1} such that, restricted to M^n the vectors $e_1, e_2, \dots, e_n$ are tangent to M^n; the vector e_{n+1} is normal to M^n. We shall make use of the following convention on the ranges of indices: $1 \leq i, j, k, \dots \leq n$, $1 \leq A, B, C, \dots \leq n+1$ and we shall agree that repeated indices are summed over the respective ranges.

With respect to the frame $e_1, e_2, \dots, e_n, e_{n+1}$ let us denote by h_{ij} the components of the second fundamental form of M^n and by R^m_{ijk} and K^D_{ABC} the components of the curvature tensors of M^n and N^{n+1} respectively. We call

$$H = \frac{1}{n} \sum_{i=1}^{n} h_{ii} e_{n+1} \tag{2.1}$$

the mean curvature vector of M^n. The square of the length of the second fundamental form of M^n is given by

$$S = \sum_{i,j=1}^{n} (h_{ij})^2. \tag{2.2}$$

S and $H^2 = \frac{1}{n^2}(\sum_{i=1}^n h_{ii})^2$ are independent of the choice of the orthonormal frame.

It is well known that for an arbitrary submanifold M^n of a Riemannian manifold N^{n+1} we have

$$R_{ijkl} = K_{ijkl} + h_{ik}h_{jl} - h_{il}h_{jk} \text{ (Gauss equation of } M^n\text{)}, \tag{2.3}$$

$$h_{ijk} - h_{ikj} = K^{n+1}_{ikj} = -K^{n+1}_{ijk} \text{ (Codazzi equation of } M^n\text{)}. \tag{2.4}$$

Here h_{ijk} is the covariant derivative of h_{ij}.

$$h_{ijkl} - h_{ijlk} = h_{im} R^m_{jkl} + h_{mj} R^m_{ikl}. \tag{2.5}$$

This equation is obtained from (2.15) in [1], when $\alpha = \beta = n+1$. Here h_{ijkl} is the covariant derivative of h_{ijk}.

$$K^{n+1}_{ijk;l} = K^{n+1}_{ijkl} - K^{n+1}_{i\,n+1\,k} h_{jl} - K^{n+1}_{ij\,n+1} h_{kl} + K^m_{ijk} h_{ml}. \tag{2.6}$$

This equation is obtained from (2.17) in [1]. Here $K^{n+1}_{ijk;l}$ is the restriction to M^n of the covariant derivative $K^A_{BCD;E}$ of K^A_{BCD} as a curvature tensor of N^{n+1}.

We consider K^{n+1}_{ijk} as a section of the bundle $T^\perp \otimes T^* \otimes T^* \otimes T^*$ where T is the tangent bundle $T = T(M)$, $T^* = T^*(M)$ - the cotangent bundle and $T^\perp = T^\perp(M)$ - the normal bundle. K^{n+1}_{ijkl} is the covariant derivative of K^{n+1}_{ijk} with respect to the covariant differentiation which maps a section of $T^\perp \otimes T^* \otimes T^* \otimes T^*$ into a section of $T^\perp \otimes T^* \otimes T^* \otimes T^* \otimes T^*$. K^{n+1}_{ijkl} must be distinguished from $K^{n+1}_{ijk;l}$.

Since we suppose that N^{n+1} is a locally symmetric one, then $K^A_{BCD;E} = 0$ and from (2.6) we obtain that

$$K^{n+1}_{ijkl} = K^{n+1}_{i\,n+1\,k}h_{jl} + K^{n+1}_{ij\,n+1}h_{kl} - K^m_{ijk}h_{ml}. \tag{2.7}$$

The Laplacian Δh_{ij} of the second fundamental form h of M^n is defined by

$$\Delta h_{ij} = \sum_{k=1}^{n} h_{ijkk}. \tag{2.8}$$

From (2.4) we obtain

$$\Delta h_{ij} = h_{ikjk} - K^{n+1}_{ijkk} = h_{kijk} - K^{n+1}_{ijkk}. \tag{2.9}$$

From (2.5) we have

$$h_{kijk} = h_{kikj} + h_{km}R^m_{ijk} + h_{mi}R^m_{kjk}. \tag{2.10}$$

In (2.10) replacing h_{kikj} by $h_{kkij} - K^{n+1}_{kikj}$ from (2.4) and putting it in (2.9) we obtain

$$\Delta h_{ij} = h_{kkij} - K^{n+1}_{kikj} - K^{n+1}_{ijkk} + h_{km}R^m_{ijk} + h_{mi}R^m_{kjk}. \tag{2.11}$$

Then from (2.11), (2.7) and (2.3) it follows that

$$\begin{aligned}\Delta h_{ij} = h_{kkij} - h_{kk}K^{n+1}_{ij\,n+1} - h_{ij}K^{n+1}_{k\,n+1\,k} + \\ h_{mj}R^m_{kik} + h_{mi}R^m_{kjk} + 2h_{mk}R^m_{ijk} - h_{mj}h_{mi}h_{kk} + h_{ij}h^2_{mk}.\end{aligned} \tag{2.12}$$

As M^n is minimal, i.e. $H = \frac{1}{n}\sum_{i=1}^n h_{ii}e_{n+1} = 0$, then from (2.12) we obtain for the Laplacian Δh_{ij} of the second fundamental form of a minimal hypersurface M^n of N^{n+1}

$$\Delta h_{ij} = -h_{ij}K^{n+1}_{k\,n+1\,k} + h_{mj}R^m_{kik} + h_{mi}R^m_{kjk} + 2h_{mk}R^m_{ijk}) + h_{ij}h^2_{mk}. \tag{2.13}$$

3. Minimal hypersurfaces of a special class of a locally symmetric manifolds.

In this section we assume that the ambient space N^{n+1} is a locally symmetric one with sectional curvature K_{N_x} satisfying the condition $\delta \leq K_{N_x} \leq 1$ at all points $x \in M^n$ and $\delta > \frac{1}{2}$. We shall prove the following theorem:

Theorem 1. Let M^n be a compact minimal hypersurface of a locally symmetric manifold N^{n+1} whose sectional curvature K_{N_x} at all points $x \in M^n$ satisfies

$$\delta \leq K_{N_x} \leq 1 \; for \; \delta > \frac{1}{2}. \tag{3.1}$$

If the square of the length S of the second fundamental form of M^n satisfies the condition

$$S \leq \frac{(2\delta - 1)n}{n-1}, \tag{3.2}$$

then S is constant.

Proof. For the Laplacian ΔS of $S = \sum_{i,j=1}^{n} (h_{ij})^2$ we have

$$\frac{1}{2}\Delta S = \sum_{i,j=1}^{n} h_{ij}\Delta h_{ij} + \sum_{i,j,k=1}^{n} (\nabla h_{ij})^2. \tag{3.3}$$

We replace in (3.3) Δh_{ij} from (2.13) and obtain

$$\frac{1}{2}\Delta S \geq -(h_{ij})^2 K^{n+1}_{k\,n+1\,k} + 2(h_{ij}h_{mj}R^m_{kik} + h_{ij}h_{mk}R^m_{ijk}) + (h_{ij})^2 h^2_{mk}. \tag{3.4}$$

We shall prove that the right side of (3.4) is non-negative. If we denote by λ_i the eigenvalues of the matrix (h_{ij}) of the second from of M^n, then Yau's formula (10.9) from [2] gives that

$$\begin{aligned} 2(h_{ij}h_{mj}R^m_{kik} + h_{ij}h_{mk}R^m_{ijk}) &= \sum_{ij}^{n} (\lambda_i - \lambda_j)^2 R_{ijij} \\ &\geq \sum_{ij}^{n} (\lambda_i - \lambda_j)^2 K = 2nKS \end{aligned} \tag{3.5}$$

where $K(x)$ is a function which assigns to each point $x \in M^n$ the infinimum of the sectional curvature of M^n at that point.

For the sectional curvature $K_M(\sigma)$ of an arbitrary submanifold M^n of a Riemannian manifold N^{n+p} at a point $x \in M^n$ for the plane σ we have from [4] the following estimate:

$$K_M(\sigma) \geq K_N(\sigma) + \frac{1}{2}(\frac{n^2H^2}{n-1} - S) \tag{3.6}$$

where $K_N(\sigma)$ is the sectional curvature of N^{n+p} at the point $x \in M^n$. Because of (3.1) and $H = 0$ the inequality (3.6) takes the form

$$K_M(\sigma) \geq \delta - \frac{1}{2}S. \tag{3.7}$$

Taking into account (3.7) for the left side of (3.5) we obtain the following lower estimate:

$$2(h_{ij}h_{mj}R^m_{kik} + h_{ij}h_{mk}R^m_{ijk}) \geq 2n\delta S - nS^2. \tag{3.8}$$

Because of (3.1) we have

$$K^{n+1}_{k\,n+1\,k} \leq n. \tag{3.9}$$

For ΔS we obtain from (3.4) because of (3.8), (3.9) and (2.2) that

$$\frac{1}{2}\Delta S \geq S[n(2\delta - 1) - (n-1)S]. \tag{3.10}$$

From (3.10) in view of (3.2) we obtain that

$$\frac{1}{2}\Delta S \geq 0. \tag{3.11}$$

Next, from the Hopf principle it follows that $S = const$. Hence the theorem is proved.

Theorem 2. Let M^n be a complete, connected minimal hypersurface of a locally symmetric manifold N^{n+1} whose sectional curvature K_{N_x} at all points $x \in M^n$ satisfies (3.1). If the square of the length S of the second fundamental form of M^n satisfies

$$S \leq 2\delta - 1, \tag{3.12}$$

then M^n is totally geodesic.

Proof. From (3.10) and (3.12) it follows that

$$\frac{1}{2}\Delta S \geq S(2\delta - 1). \tag{3.13}$$

Since S is boundet above and the sectional curvature of M^n is bounded below, we claim that $S = 0$ everywhere on M^n. In fact, if for some point $p \in M^n$ we had $S(p) = a > 0$, then from (3.13) we should have $\frac{1}{2}\Delta S \geq a(2\delta - 1) = const. > 0$, and thus for all points $q \in M^n$ for which $S(q) \geq S(p)$ we ought have $\frac{1}{2}\Delta S(q) \geq a(2\delta - 1) > 0$ which contradicts Omori's theorem A' in [3]. So, $S = 0$ everywhere on M^n which means that M^n is totally geodesic. Hence the theorem is proved.

References.

[1] Chern, S.S., M. do Carmo and Kobayashi S.: Minimal submanifolds of a sphere with second fundamental form of constant length. Functional Analysis and Related Fields, 59-75 (1970).

[2] Yau, S. T.: Submanifolds with constant mean curvature II. Amer. J. of Math. 97, No. 1, 76-100 (1975).

[3] Omori, H.: Isometric immersions of Riemannian manifolds. J. Math. Soc. Japan 19 , 205-214 (1967).

[4] Hineva, S. T.: Submanifolds and the second fundamental tensor. Lecture Notes in Math. 1156, 194-203 (1984).

This paper is in final form and no version will appear elsewhere.

Faculty of Mathematics and Informatics
University of Sofia
Anton Ivanov Str. 5
1126 Sofia, Bulgaria

The Spectral Geometry of the Laplacian and the Conformal Laplacian for Manifolds with Boundary

Novica Blažić
Neda Bokan
Peter Gilkey

Abstract: We study the spectral geometry of the Laplacian with Dirichlet and Neumann boundary conditions and the spectral geometry of the conformal Laplacian with Dirichlet and Robin boundary conditions. We show in §1 geometric properties of the boundary such as totally geodesic boundary, constant mean curvature, and totally umbillic are spectrally determined. In §2, we expand the invariants of the heat equation on a small geodesic ball in a power series in the radius. We characterize Einstein, conformally flat, and constant sectional curvature manifolds by the spectral geometry of their geodesic balls. Also, some characterizations are obtained for the rank 1 symmetric spaces S^n, CP^n, QP^n, CaP^2 and their noncompact duals. MOS subject classification: 58G25

§0 Heat Equation Asymptotics

Let M be a compact Riemannian manifold of dimension m with smooth boundary ∂M and let $\Delta = \delta d$ be the scalar Laplacian on M. We suppose $m > 1$. Let ∇ be the Levi-Civita connection of M. We adopt the convention that indices $\{i, j, ...\}$ range from 1 through m and index a local orthonormal frame $\{e_i\}$ for TM. We shall sum over repeated indices. The curvature tensor R_{ijkl} is defined by:

$$R_{ijkl} = ((\nabla_{e_i}\nabla_{e_j} - \nabla_{e_j}\nabla_{e_i} - \nabla_{[e_i,e_j]})e_k, e_l); \tag{0.1}$$

$R_{1212} = -1$ on $S^2 \subseteq \mathbf{R}^3$. Let $\rho_{ij} = -R_{ikjk}$ and $\tau = \rho_{ii} = -R_{ijij}$ define the Ricci tensor ρ and scalar curvature τ.

The Laplacian Δ is not conformally invariant if $m > 2$. The conformal Laplacian $\tilde{\Delta}$ is defined by adjusting Δ by a suitable multiple of the scalar curvature;

$$\tilde{\Delta} = \Delta + \tfrac{(m-2)}{4(m-1)}\tau. \tag{0.2}$$

This paper is in final form and no version of it will be submited for publication elsewhere.

Research of P. Gilkey partially supported by the NSF and NSA.

Research of N. Blažić and N. Bokan partially supported by the conference "Global Differential Geometry and Global Analysis", TU Berlin, June 15–20, 1990.

Let $g = e^{2f} \cdot g_0$ be a conformally equivalent metric. Then

$$\tilde{\Delta}(g) = e^{-2f} e^{-\alpha f} \tilde{\Delta}\,(g_0) e^{\alpha f} \tag{0.3}$$

where $\alpha = \frac{1}{2}(m-2)$; see for example Branson et al [BO].

Let D be either the Laplacian Δ or the conformal Laplacian $\tilde{\Delta}$ henceforth. Since $\partial M \neq \emptyset$, we must impose suitable boundary conditions to ensure D has point spectrum. Let $f \in C^\infty(M)$. $B_D f = f|_{\partial M}$ defines *Dirichlet boundary conditions*. Let $\partial_\nu \in T(\partial M)$ be the inward unit normal vector field. If $S \in C^\infty(M)$ is an auxilary real function, $B_N^S f = (\partial_\nu + S)|_{\partial M}$ defines *modified Neumann boundary conditions;* let $B_N = B_N^0$ define *pure Neumann boundary conditions.*

The boundary conditions B_D for $\tilde{\Delta}$ are conformally invariant; the boundary conditions B_N are not. The second fundamental form plays an important role in the discussion. Near the boundary choose the frame field so e_m is the extension of the inward unit normal to a collared neighborhood by parallel transport along the inward pointing geodesic rays. Let indices $\{a, b, ...\}$ range from 1 thru $m-1$ and index a local orthonormal frame $\{e_a\}$ for $T(\partial M)$. Let L_{ab} be the second fundamental form;

$$L_{ab} = (\nabla_{e_a} e_b, e_m). \tag{0.4}$$

Let

$$c(m) = \tfrac{(m-2)}{2(m-1)} \text{ and } B_{Ro} f = (\partial_\nu f - c(m) L_{aa} f)|_{\partial M}. \tag{0.5}$$

define *Robin boundary conditions;*

$$B_{Ro}(g) = e^{-f} e^{-\alpha f} B_{Ro}(g_0) e^{\alpha f}. \tag{0.6}$$

(0.3) and (0.6) can be recast in the following language. Let L be the trivial line bundle of densities of order α. Then $\tilde{\Delta}$ and B_{Ro} extend to operators $\hat{\Delta}$ and $\hat{B}_{Ro}$ on $C^\infty(L)$ which satisfy:

$$\hat{\Delta}(g) = e^{-2f} \hat{\Delta}(g_0) \text{ and } \hat{B}_{Ro}(g) = e^{-f} \hat{B}_{Ro}(g_0). \tag{0.7}$$

Let $D = \Delta$ or $D = \tilde{\Delta}$. If $D = \Delta$, let $B = B_N$ or $B = B_D$; if $D = \tilde{\Delta}$, let $B = B_{Ro}$ or $B = B_D$. Let D_B be D acting on $\{f \in C^\infty(M) : Bf = 0\}$. D_B is self-adjoint. Let $\text{spec}(D, B)_M = \{\lambda_i\}$ be the spectrum of D_B. The λ_i are real and as $i \to \infty$, $\lambda_i \to \infty$. The L^2 trace of the heat kernel

$$Tr_{L^2}\{e^{-tD_B}\} = \Sigma_i\, e^{-t\lambda_i} \tag{0.8}$$

defines an analytic function of t for $t > 0$. As $t \to 0^+$ there is an asymptotic expansion

$$Tr_{L^2}\{e^{-tD_B}\} \simeq \Sigma_{n \geq 0}\, a_n(D, B) t^{(n-m)/2}. \tag{0.9}$$

The $a_n(D, B)$ are spectral invariants.

Remark: In the literature, this asymptotic sum is often reindexed since the odd terms a_{2i+1} vanish if $\partial M = \emptyset$. We do not adopt this convention as we are considering manifolds with boundary.

The invariants $a_n(D,B)$ are given by integrals of local geometric quantities. If $\partial M=\emptyset$, $a_n(D,B)=a_n(D)$ can be computed in terms of curvature. These invariants have been used by many authors to relate the spectrum of the Laplacian to the geometry of the underlying manifold. Let

$$E=\begin{cases}0 & \text{if } D=\Delta\\ -\frac{(m-2)}{4(m-1)}\tau & \text{if } D=\tilde{\Delta}\end{cases} \tag{0.10}$$

We define the following positive normalizing constants:

$$c_0(m)=(4\pi)^{-m/2} \tag{0.11}$$
$$c_1(m)=4^{-1}(4\pi)^{-(m-1)/2} \tag{0.12}$$
$$c_2(m)=6^{-1}(4\pi)^{-m/2} \tag{0.13}$$
$$c_3(m)=384^{-1}(4\pi)^{-(m-1)/2} \tag{0.14}$$
$$c_4(m)=360^{-1}(4\pi)^{-m/2} \tag{0.15}$$

Let dV be the Riemannian measure on M. We refer to [G, Theorem 4.1] for the proof of:

Theorem 0.1: *Let $\partial M=\emptyset$. $a_n(D)=0$ for n odd.*

(a) $a_0(D)=c_0(m)\int_M dV$.
(b) $a_2(D)=c_2(m)\int_M(\tau+6E)dV$.
(c) $a_4(D)=c_4(m)\int_M(5\tau^2-2\rho^2+2R^2+180E^2+60E\tau)dV$.
(d) $a_6(D)=(4\pi)^{-m/2}\int_M\{45360^{-1}(-142(\nabla\tau)^2-26(\nabla\rho^2)-7(\nabla R)^2+35\tau^3$
$-42\tau\rho^2+42\tau R^2-36\rho_{ij}\rho_{jk}\rho_{ki}-20\rho_{ij}\rho_{kl}R_{ikjl}-8\rho_{ij}R_{ikln}R_{jkln}$
$-24R_{ijkl}R_{ijnp}R_{klnp})$
$+360^{-1}(-30E_{;i}E_{;i}+60E^3-12E_{;k}\tau_{;k}+5E\tau^2-2E\rho^2+2ER^2)\}dV$.

Remark: $a_6(\Delta)$ has also been computed by Sakai [Sa, Theorem 4.2]. Avramidi [Av, see §3.4] and Amsterdamski et al [ABC, see §III] have computed $a_8(D)$; it has formidable combinatorial complexity.

If $\partial M\neq\emptyset$, there are additional boundary contributions and the $a_n(D,B)$ do not vanish for n odd. If θ is a tensor field, let $\theta_{;i}$ be the components of $\nabla\theta$; similarly $\theta_{;ij}$ are the components of $\nabla^2\theta$ and so forth. Define E, χ, Π_D, Π_N, and S by:

(D,B)	E	χ	Π_D	Π_N	S
(Δ,B_D)	0	-1	1	0	0
(Δ,B_N)	0	1	0	1	0
$(\tilde{\Delta},B_D)$	$-\frac{(m-2)}{4(m-1)}\tau$	-1	1	0	0
$(\tilde{\Delta},B_{Ro})$	$-\frac{(m-2)}{4(m-1)}\tau$	1	0	1	$-\frac{(m-2)}{2(m-1)}L_{aa}$

Let the constants c_i be as defined in (0.11-0.15). Let dv be the Riemannian measure on ∂M. We refer to [BG, Theorem 7.2] for a proof using functorial methods of:

Theorem 0.2:

(a) $a_0(D,B) = c_0(m)\int_M dV$.

(b) $a_1(D,B) = c_1(m)\int_{\partial M}\chi dv$.

(c) $a_2(D,B) = c_2(m)\{\int_M(6E+\tau)dV + \int_{\partial M}(2L_{aa}+12S)dv\}$.

(d) $a_3(D,B) = c_3(m)\int_{\partial M}\{96\chi E + 16\chi\tau + 96SL_{aa} + 192S^2$
$+(13\Pi_N - 7\Pi_D)L_{aa}L_{bb} + (2\Pi_N + 10\Pi_D)L_{ab}L_{ab} - 8\chi\rho_{mm}\}dv$.

(e) $a_4(D,B) = c_4(m)\{\int_M(60E_{;kk} + 12\tau_{;kk} + 60\tau E + 180E^2 + 5\tau^2 - 2\rho^2 + 2R^2)dV$
$+\int_{\partial M}\{(240\Pi_N - 120\Pi_D)E_{;m} + 120S\tau + 144SL_{aa}L_{bb}$
$+(42\Pi_N - 18\Pi_D)\tau_{;m} + 120EL_{aa} + 20\tau L_{aa} + 720SE + 480S^3$
$+21^{-1}(280\Pi_N + 40\Pi_D)L_{aa}L_{bb}L_{cc} + 21^{-1}(168\Pi_N - 264\Pi_D)L_{ab}L_{ab}L_{cc}$
$-12R_{ambm}L_{ab} + 48SL_{ab}L_{ab} + 4R_{amam}L_{bb} + 480S^2L_{aa}$
$+21^{-1}(224\Pi_N + 320\Pi_D)L_{ab}L_{bc}L_{ac} + 4R_{abcb}L_{ac}\}dv$.

Remark: The formulae of [BG] deal with arbitrary vector valued second order operators with leading symbol given by the metric tensor and boundary conditions which are mixtures of Dirichlet and Neumann conditions; we have specialized these formulae to the case of Δ and $\tilde{\Delta}$. McKean et al [MS, see page 45] in 1967 computed $a_n(\Delta,B)$ for $n=0,1$. Kennedy et al [KCD, see page 378] in 1980 computed $a_n(\Delta,B)$ and $a_n(\tilde{\Delta},B)$ for $n=2,3$. Moss and Dowker [MD] computed $a_4(\Delta,B)$ and $a_4(\tilde{\Delta},B)$ independently of the computation in [BG].

In §1,we will show the spectral geometry determines if the boundary is totally geodesic, is totally umbillic, or is constant mean curvature. In §2, we will expand the heat equation invariants $a_n(D,B)$ for small geodesic balls $B_r(x)$ in a power series in r. This leads to characterizations of Einstein, conformally flat, and constant curvature manifolds in terms of spectral geometry.

§1 Spectral geometry of the boundary

We say that ∂M is totally umbillic if there exists $f \in C^\infty(\partial M)$ so $L_{ab} = f(x)g_{ab}$. We say ∂M is totally geodesic if $L \equiv 0$ on ∂M. We say that ∂M has constant mean curvature if $L_{aa} \equiv c$ on ∂M. These properties are spectrally determined. We begin our study with a technical Lemma:

Lemma 1.1: *Let* $\delta(L) = (m-1)L_{ab}L_{ab} - L_{aa}L_{bb}$.

(a) $\delta(L) \geq 0$.

(b) $\delta(L) \equiv 0$ *if and only if* ∂M *is totally umbillic.*

Proof: If x and y are real numbers, then $xy \leq \frac{1}{2}(x^2 + y^2)$. Equality holds if and only if $x = y$. Consequently

$$L_{aa}L_{bb} \leq (m-1)L_{aa}^2 \leq (m-1)L_{ab}L_{ab} \tag{1.1}$$

Equality holds if and only if $L_{ab} = 0$ for $a \neq b$ and if all elements on the diagonal are equal. This means ∂M is totally umbillic. ■

Theorem 1.2:

(a) *Assume* $spec(\Delta, B)_{M_1} = spec(\Delta, B)_{M_2}$ *for* $B = B_D$ *and* $B = B_N$. *If* ∂M_1 *is totally geodesic, then* ∂M_2 *is totally geodesic.*

(b) *Let* $m \neq 3, 4, 5$. *Assume* $spec(\tilde{\Delta}, B)_{M_1} = spec(\tilde{\Delta}, B)_{M_2}$ *for* $B = B_D$ *and* $B = B_{Ro}$. *If* ∂M_1 *is totally geodesic, then* ∂M_2 *is totally geodesic.*

(c) *Let* $m = 3$ *or* $m = 5$. *Assume* $spec(\tilde{\Delta}, B_D)_{M_1} = spec(\tilde{\Delta}, B_D)_{M_2}$ *for* $B = B_D$ *and* $B = B_{Ro}$. *If* ∂M_1 *is totally umbillic, then* ∂M_2 *is totally umbillic.*

Proof: Let $c(m) = \frac{(m-2)}{2(m-1)}$ be as in (0.5). By Theorem 0.2, $\exists c_3 > 0$ so

$$a_3(\Delta, B_D) + a_3(\Delta, B_N) = c_3 \int_{\partial M} \{6L_{aa}L_{bb} + 12L_{ab}L_{ab}\} dv \tag{1.2}$$

$$\begin{aligned} a_3(\tilde{\Delta}, B_D) + a_3(\tilde{\Delta}, B_{Ro}) = c_3 \int_{\partial M} \{ &(6 - 96c(m) \\ &+ 192c(m)^2)L_{aa}L_{bb} + 12L_{ab}L_{ab}\} dv. \end{aligned} \tag{1.3}$$

The following assertions are equivalent:

(1) ∂M is totally geodesic.

(2) $L \equiv 0$.

(3) $\int_M L_{ab}L_{ab} dv = 0$.

(4) $\int_M (6L_{aa}L_{bb} + 12L_{ab}L_{ab}) dv = 0$.

As (4) is spectrally determined by (1.2), (a) follows.

We use (1.3) to prove (b) and (c). The coefficient of $L_{aa}L_{bb}$ is positive for $m = 2$ or $m \geq 8$ so (b) follows using the same argument in these cases. Let $d(m) = 6 - 96c(m) + 192c(m)^2$. By Lemma 1.1, for $3 \leq m \leq 7$,

$$\int_{\partial M} (d(m)L_{aa}L_{bb} + 12L_{ab}L_{ab}) dv \geq (12 + (m-1) \cdot d(m)) \int_{\partial M} L_{ab}L_{ab} dv. \tag{1.4}$$

If $m = 6$, then $12 + (m-1) \cdot d(m) = 3.6$; if $m = 7$, then $12 + (m-1) \cdot d(m) = 8$. Since this coefficient is positive, (1.3) is non-negative and vanishes if and only if $L_{ab} \equiv 0$. This proves (b). If $m = 3$ or if $m = 5$, then this coefficient is zero. Thus (1.4) is non-negative. Equality holds if and only if equality holds in Lemma 1.1 or equivalently if ∂M is totally umbillic. (c) now follows. ■

Theorem 1.2: *Let $m \geq 3$. Assume $spec(\Delta, B)_{M_1} = spec(\Delta, B)_{M_2}$ for $B = B_D$ and $B = B_N$ and that $spec(\tilde{\Delta}, B)_{M_1} = spec(\tilde{\Delta}, B)_{M_2}$ for $B = B_D$ and $B = B_{Ro}$.*

(a) ∂M_1 has constant mean curvature c if and only if ∂M_2 has constant mean curvature c.

(b) ∂M_1 is totally umbillic if and only if ∂M_2 is totally umbillic.

Proof: Let $c(m) = \frac{(m-2)}{2(m-1)}$. By Theorem 0.2, $\exists c_i > 0$ so

$$\begin{aligned} a_1(\Delta, B_N) &= c_1 \cdot \mathrm{vol}(\partial M) && (1.5)\\ a_2(\tilde{\Delta}, B_{Ro}) - a_2(\tilde{\Delta}, B_D) &= c_2 \textstyle\int_{\partial M}(-12c(m)L_{aa})dv && (1.6)\\ a_3(\Delta, B_D) + a_3(\Delta, B_N) &= c_3 \textstyle\int_{\partial M}(6L_{aa}L_{bb} + 12L_{ab}L_{ab})dv && (1.7)\\ a_3(\tilde{\Delta}, B_D) + a_3(\tilde{\Delta}, B_{Ro}) &= c_3 \textstyle\int_{\partial M}\{(6 - 96c(m) + 192c(m)^2)L_{aa}L_{bb} && (1.8)\\ &\quad + 12L_{ab}L_{ab}\}dv. \end{aligned}$$

Since $m \geq 3$, $c(m) \in (0, \frac{1}{2})$. We take linear combinations of (1.5)-(1.8) to see

$$\{\mathrm{vol}(\partial M),\ \textstyle\int_{\partial M} L_{aa}dv,\ \int_{\partial M} L_{aa}L_{bb}dv,\ \text{and}\ \int_{\partial M} L_{ab}L_{ab}dv\}$$

are spectral invariants. The following assertions are equivalent:

(1) M has constant mean curvature c

(2) $L_{aa} \equiv c$

(3) $\int_{\partial M}(L_{aa} - c)^2 dv = 0$.

(4) $\int_{\partial M} L_{aa}L_{bb}dv - 2c\int_{\partial M} L_{aa}dv + c^2 = 0$.

(a) now follows as (4) is a spectral invariant. Since $\delta(M)$ is a spectral invariant, we use Lemma 1.1 to prove (b). ■

§2 Heat equations on small geodesic balls

Let M be a compact Riemannian manifold without boundary and let $\iota(M)$ be the injectivity radius of M. For $r < \iota(M)$, let $B(r, x)$ be the geodesic ball of radius r centered at a point $x \in M$. Let $S(r, x) = \partial(B(r, x))$ be the corresponding geodesic sphere. In this section, we expand $a_n(D, B)_{B(r,x)}$ in an asymptotic series in the parameter r where (D, B) is one of the 4 operators and boundary conditions considered previously. This leads to a calculation of "asymptotics of the asymptotics" in Theorems 2.4-2.8 which are of interest in their own right. We apply these results to derive various conclusions in spectral geometry at the end of §2. We remark that Karp and Pinsky [KP] have generalized results of Levy-Bruhl [Le] and give the first three terms in the asymptotic expansion of the lowest eigenvalue $\lambda_1(\Delta, B_D, r)$.

Fix $x \in M$. Let $S = S(x) \subseteq TM_x$ be the unit sphere. The innerproduct on TM_x defines a natural volume element $d\xi$ on S; $r^{m-1}d\xi \wedge dr$ is the natural volume element of TM_x. In Lemma 2.1, we integrate certain tensorial expressions which are defined on TM_x over S with respect to the measure $d\xi$.

The exponential map defines geodesic polar coordinates (r,ξ) on M for $\xi \in S$ and for $r \in [0, \iota(M))$. The geodesic sphere $S(x,r)$ is the image of $r \cdot S$ under the exponential map and the outward pointing normal can be identified with either dr or with ξ. Let $\mathrm{dvol}(r,\xi)$ be the pull back of the Riemannian volume element dV on M to TM_x. Express

$$\mathrm{dvol}(r,\xi) = r^{m-1} d\theta(r,\xi) \wedge dr; \tag{2.1}$$

where $d\theta(r,\xi) = \omega(r,\xi)d\xi$ is another volume element on S. In Lemma 2.2, we expand integrals of invariants over $S(x,r)$ in a power series in r with coefficients depending on x. Lemma 2.3 is a similar expansion for the integral over $D(x,r)$.

Lemmas 2.1 and 2.2 depend heavily on the results of Chen and Vanhecke [CVH]. We refer to [CVH, Theorem 6.4] for the proof of Lemma 2.1. Let

$$v(m) = \textstyle\int_S d\xi = m \cdot \pi^{m/2} \cdot \Gamma(\frac{1}{2}m + 1)^{-1} \tag{2.2}$$
$$\alpha(m) = v(m)/\{6m\} \tag{2.3}$$
$$\beta(m) = v(m)/\{360m(m+2)\} \tag{2.4}$$

Lemma 2.1:

(a) $\int_S \rho_{\xi\xi} d\xi = \frac{v(m)}{m}\tau.$

(b) $\int_S \rho_{\xi\xi;\xi} d\xi = \int_S \tau_{;\xi} d\xi = 0.$

(c) $\int_S \tau_{;\xi\xi} d\xi = \frac{-v(m)}{m}\Delta\tau.$

(d) $\int_S \rho_{\xi\xi;\xi\xi} d\xi = \frac{-2v(m)}{m(m+2)}\Delta\tau.$

(e) $\int_S \rho_{\xi\xi}\rho_{\xi\xi} d\xi = \frac{v(m)}{m(m+2)}(\tau^2 + 2\rho^2).$

(f) $\int_S R_{\xi a\xi b}R_{\xi a\xi b} d\xi = \frac{v(m)}{2m(m+2)}(2\rho^2 + 3R^2).$

Lemma 2.1 deal with integrals in the tangent space; Lemma 2.2 deals with corresponding integrals over the manifold.

Lemma 2.2:

(a) $\int_{S(r,x)} d\theta(r,\xi) = v(m)r^{m-1} - r^{m+1}\alpha(m)\tau + r^{m+3}\beta(m)(18\Delta\tau + 5\tau^2 + 8\rho^2$
$-3R^2) + O(r^{m+5}).$

(b) $\int_{S(r,x)} L_{aa} d\theta(r,\xi) = v(m)r^{m-2}(m-1) - r^m\alpha(m)(m+1)\tau$
$+r^{m+2}\beta(m)(m+3)(18\Delta\tau + 5\tau^2 + 8\rho^2 - 3R^2) + O(r^{m+4})\}.$

(c) $\int_{S(r,x)} \tau d\theta(r,\xi) = r^{m-1}\alpha(m)6m\tau - r^{m+1}\beta(m)(m+2)(180\Delta\tau + 60\tau^2)$
$+O(r^{m+3}).$

(d) $\int_{S(r,x)} \rho_{\xi\xi} d\theta(r,\xi) = r^{m-1}\alpha(m)6\tau - r^{m+1}\beta(m)(360\Delta\tau + 60\tau^2 + 120\rho^2)$
$+O(r^{m+3}).$

(e) $\int_{S(r,x)} L_{aa}L_{bb} d\theta(r,\xi) = v(m)r^{m-3}(m-1)^2 - r^{m-1}\alpha(m)(m-1)(m+3)\tau$
$+r^{m+1}\beta(m)\{18(m^2+6m-7)\Delta\tau + 5(m^2+6m+1)\tau^2$
$+8(m^2+6m+3)\rho^2 - 3(m^2+6m-7)R^2 + O(r^{m+3}).$

(f) $\int_{S(r,x)} L_{ab}L_{ab} d\theta(r,\xi) = v(m)r^{m-3}(m-1) - r^{m-1}\alpha(m)(m+3)\tau$
$+r^{m+1}\beta(m) \cdot \{18(m+7)\Delta\tau + 5(m+7)\tau^2 + 8(m+12)\rho^2$
$-3(m-13)R^2\} + O(r^{m+3}).$

Remark: The formulas (a), (b) and (f) are known (see [Gr] and [GV]).

Proof: By [CVH, Corollary 2.4],

$$\begin{aligned} d\theta(r,\xi) =&\{r^{m-1} - \tfrac{r^{m+1}}{6}\rho_{\xi\xi} - \tfrac{r^2}{12}\rho_{\xi\xi;\xi} + \tfrac{r^{m+3}}{360}(-9\rho_{\xi\xi;\xi\xi} \\ &+ 5\rho_{\xi\xi}\rho_{\xi\xi} - 2R_{\xi a\xi b}R_{\xi a\xi b}) + O(r^{m+4})\}d\xi. \end{aligned} \tag{2.5}$$

We integrate (2.5) using Lemma 2.1 to prove (a); the coefficient of r^{m+4} is odd in ξ and hence integrates to 0 so the error is $O(r^{m+5})$ not $O(r^{m+4})$. By [CVH, Theorem 3.1],

$$\begin{aligned} L_{ab}(r,\xi) =&r^{-1}\delta_{ab} + \tfrac{r}{3}R_{\xi a\xi b} + \tfrac{r^2}{4}R_{\xi a\xi b;\xi} + \tfrac{r^3}{360}(36R_{\xi a\xi b;\xi\xi} \\ &- 8R_{\xi a\xi s}R_{\xi b\xi s}) + O(r^4). \end{aligned} \tag{2.6}$$

We set $a = b$ and combine (2.5) and (2.6) to compute:

$$\begin{aligned} L_{aa}d\theta(r,\xi) =&\{(m-1)r^{m-2} - \tfrac{r^m(m+1)}{6}\rho_{\xi\xi} - \tfrac{r^{m+1}(m+2)}{12}\rho_{\xi\xi;\xi} \\ &+ \tfrac{r^{m+2}(m+3)}{360}(-9\rho_{\xi\xi;\xi\xi} - 2R_{\xi a\xi s}R_{\xi a\xi s} + 5\rho_{\xi\xi}\rho_{\xi\xi}) \\ &+ O(r^{m+3})\}d\xi. \end{aligned} \tag{2.7}$$

We integrate (2.7) using Lemma 2.1 to prove (b). By [CVH, Lemma 4.1],

$$\tau(r,\xi) =\tau + r\tau_{;\xi} + \tfrac{r^2}{2}\tau_{;\xi\xi} + O(r^3) \tag{2.8}$$

$$\rho_{ij}(r,\xi) =\rho_{ij} + r\rho_{ij;\xi} + \tfrac{r^2}{2}\rho_{ij;\xi\xi} + O(r^3). \tag{2.9}$$

We combine (2.5), (2.8), and (2.9) to compute:

$$\tau d\theta(r,\xi) =\{r^{m-1}\tau + r^m\tau_{;\xi} + \tfrac{r^{m+1}}{360}(180\tau_{;\xi\xi} - 60\tau\rho_{\xi\xi}) + O(r^{m+2})\}d\xi \tag{2.10}$$

$$\begin{aligned} \rho_{\xi\xi}d\theta(r,\xi) =&\{r^{m-1}\rho_{\xi\xi} + r^m\rho_{\xi\xi;\xi} + \tfrac{r^{m+1}}{360}(180\rho_{\xi\xi;\xi\xi} - 60\rho_{\xi\xi}\rho_{\xi\xi}) \\ &+ O(r^{m+2})\}d\xi. \end{aligned} \tag{2.11}$$

We integrate (2.10) and (2.11) using Lemma 2.1 to prove (c) and (d); we integrate (c) with respect to r to prove (e). We use (2.5) and (2.6) to compute

$$\begin{aligned} L_{ab}L_{cd}(r,\xi) =&r^{-2}\delta_{ab}\delta_{cd} + \tfrac{1}{3}\{R_{\xi a\xi b}\delta_{cd} + \delta_{ab}R_{\xi c\xi d}\} \\ &+ \tfrac{r}{4}\{R_{\xi a\xi b;\xi}\delta_{cd} + R_{\xi c\xi d;\xi}\delta_{ab}\} \\ &+ \tfrac{r^2}{360}(36R_{\xi a\xi b;\xi\xi}\delta_{cd} + 36\delta_{ab}R_{\xi c\xi d;\xi} - 8R_{\xi a\xi s}R_{\xi b\xi s}\delta_{cd} \\ &- 8\delta_{ab}R_{\xi c\xi s}R_{\xi d\xi s} + 40R_{\xi a\xi b}R_{\xi c\xi d}) + O(r^3) \end{aligned} \tag{2.12}$$

$$\begin{aligned} L_{ab}L_{cd}d\theta(r,\xi) =&\{r^{m-3}\delta_{ab}\delta_{cd} - \tfrac{r^{m-1}}{6}(-2R_{\xi a\xi b}\delta_{cd} \\ &- 2\delta_{ab}R_{\xi c\xi d} + \rho_{\xi\xi}\delta_{ab}\delta_{cd}) \\ &- \tfrac{r^m}{12}(-3R_{\xi a\xi b;\xi}\delta_{cd} - 3R_{\xi c\xi d}\delta_a\delta_b + \rho_{\xi\xi;\xi}\delta_{ab}\delta_{cd}) \\ &+ \tfrac{r^{m+1}}{360}(36R_{\xi a\xi b;\xi\xi}\delta_{cd} + 36\delta_{ab}R_{\xi c\xi d;\xi\xi} - 8R_{\xi a\xi s}R_{\xi b\xi s}\delta_{cd} \end{aligned} \tag{2.13}$$

$$- 8\delta_{ab}R_{\xi c\xi s}R_{\xi d\xi s} + 40R_{\xi a\xi b}R_{\xi c\xi d} - 20R_{\xi a\xi b}\delta_{cd}\rho_{\xi\xi}$$
$$- 20\delta_{ab}R_{\xi c\xi d}\rho_{\xi\xi} - 9\rho_{\xi\xi;\xi\xi}\delta_{ab}\delta_{cd} + 5\rho_{\xi\xi}\rho_{\xi\xi}\delta_{ab}\delta_{cd}$$
$$- 2R_{\xi u\xi v}R_{\xi u\xi v}\delta_{ab}\delta_{cd}) + O(r^{m+2})\}d\xi$$

$$L_{aa}L_{bb}d\theta(r,\xi) = \{r^{m-3}(m-1)^2 - \tfrac{r^{m-1}(m-1)(m+3)}{6}\rho_{\xi\xi} \tag{2.14}$$
$$- \tfrac{r^m(m-1)(m+5)}{12}\rho_{\xi\xi;\xi} + \tfrac{r^{m+1}}{360}(-9(m-1)(m+7)\rho_{\xi\xi;\xi\xi}$$
$$- 2(m-1)(m+7)R_{\xi a\xi b}R_{\xi a\xi b}$$
$$+ 5(m^2+6m+1)\rho_{\xi\xi}\rho_{\xi\xi}) + O(r^{m+2})\}d\xi$$

$$L_{ac}L_{ac}d\theta(r,\xi) = \{r^{m-3}(m-1) - \tfrac{r^{m-1}(m+3)}{6}\rho_{\xi\xi} - \tfrac{r^m}{12}(m+5)\rho_{\xi\xi;\xi} \tag{2.15}$$
$$+ \tfrac{r^{m+1}}{360}(-9(m+7)\rho_{\xi\xi;\xi\xi} + 2(13-m)R_{\xi u\xi v}R_{\xi u\xi v}$$
$$+ 5(m+7)\rho_{\xi\xi}\rho_{\xi\xi} + O(r^{m+2})\}d\xi.$$

We integrate (2.14) and (2.15) using Lemma 2.1 to complete the proof. ■

We integrate Lemma 2.2 to prove:

Lemma 2.3:

(a) $\int_{D(r,x)} d\theta(r,\xi) = \frac{r^m v(m)}{m} - \frac{r^{m+2}\alpha(m)}{(m+2)}\tau + \frac{r^{m+4}\beta(m)}{(m+4)}(18\Delta\tau + 5\tau^2 + 8\rho^2 - 3R^2)$
$+O(r^{m+6})$.

(b) $\int_{D(r,x)} \tau(r,\xi)d\theta(r,\xi)dr = r^m\alpha(m)6\tau - r^{m+2}\beta(m)(180\Delta\tau + 60\tau^2) + O(r^{m+4})$.

Remark: The expansion (a) is contained in [Gr] and [GV].

Let χ and c_i be as defined in §1;

$c_0(m) = (4\pi)^{-m/2}$	$\chi(\Delta, B_D) = -1$
$c_1(m) = 4^{-1}(4\pi)^{-(m-1)/2}$	$\chi(\tilde{\Delta}, B_D) = -1$
$c_2(m) = 6^{-1}(4\pi)^{-m/2}$	$\chi(\Delta, B_N) = 1$
$c_3(m) = 384^{-1}(4\pi)^{-(m-1)/2}$	$\chi(\tilde{\Delta}, B_{Ro}) = 1$

Let $a_n(D,B,r)$ be the invariants of the heat equation $a_n(D,B)$ on the geodesic ball $D(r)$ about some point $x \in M$. We combine Theorem 0.2 with Lemmas 2.2 and 2.3 to compute the 'asymptotics of the asymptotics'.

The expansion of $a_n(D,B,r)$ for $n = 0,1$ is essentially independent of (D,B) except for the normalizing sign χ. They contain essentially the same information.

Theorem 2.4: *Let* $(D,B) \in \{(\Delta, B_D), (\Delta, B_N), (\tilde{\Delta}, B_D), (\tilde{\Delta}, B_{Ro})\}$.

(a) $a_0(\Delta,B,r) = c_0(m)\{\frac{r^m v(m)}{m} - \frac{r^{m+2}\alpha(m)}{m+2}\tau$
$+\frac{r^{m+4}\beta(m)}{m+4}(18\Delta\tau + 5\tau^2 + 8\rho^2 - 3R^2) + O(r^{m+6})\}$.

(b) $a_1(\Delta,B,r) = \chi \cdot c_1(m)\{v(m)r^{m-1} - r^{m+1}\alpha(m)\tau$
$+r^{m+3}\beta(m)(18\Delta\tau + 5\tau^2 + 8\rho^2 - 3R^2) + O(r^{m+5})\}$.

The remaining asymptotics reflect the geometry of the operator more closely.

Theorem 2.5: *Let* $(D,B)=(\Delta,B_D)$.

$$\begin{aligned}(a)\quad & a_2(\Delta,B_D,r)=c_2(m)\{v(m)r^{m-2}2(m-1)-r^m\alpha(m)(2m-4)\tau\\ & +r^{m+2}\beta(m)(18(2m-4)\Delta(\tau)+5(2m-6)\tau^2\\ & +8(2m+6)\rho^2-3(2m+6)R^2+O(r^{m+4})\}.\\ (b)\quad & a_3(\Delta,B_D,r)=-c_3(m)\{v(m)r^{m-3}(7m^2-24m+17)\\ & -r^{m-1}\alpha(m)(7m^2-92m-3)\tau\\ & +r^{m+1}\beta(m)(18(7m^2-128m-279)\Delta\tau+5(7m^2-160m-351)\tau^2\\ & +8(7m^2+32m+21)\rho^2-3(7m^2+32m+81)R^2+O(r^{m+3})\}.\end{aligned}$$

Remark: Let $\lambda_1(\Delta,B_D,r)$ be the lowest eigenvalue of (D,B) on the ball of radius r $D(x,r)$. Karp and Pinsky [KP] generalize a result of Levy-Bruhl [Le] and expand λ_1 in an asymptotic series in r :

$$\lambda_1(\Delta,B_D,r)=b_0(m)^2r^{-2}-\tfrac{1}{6}\tau+b_2(m)r^2\{R^2-\rho^2+6\Delta\tau\}+O(r^4)$$

where the $b_0(m)$ is the smallest zero of the Bessel function $J_{(1/2)(m-2)}$ and where $b_2(m)$ is a negative constant. This provides additional spectral information in this case.

Theorem 2.6: *Let* $(D,B)=(\Delta,B_N)$.

$$\begin{aligned}(a)\quad & a_2(\Delta,B_N,r)=c_2(m)\{v(m)r^{m-2}2(m-1)-r^m\alpha(m)(2m-4)\tau\\ & +r^{m+2}\beta(m)(18(2m-4)\Delta\tau+5(2m-6)\tau^2\\ & +8(2m+6)\rho^2-3(2m+6)R^2+O(r^{m+4})\}.\\ (b)\quad & a_3(\Delta,B_N,r)=c_3(m)\{v(m)r^{m-3}(13m^2-14m+11)\\ & -r^{m-1}\alpha(m)(13m^2-68m+15)\tau\\ & +r^{m+1}\beta(m)(18(13m^2-80m-237)\Delta\tau+5(13m^2-112m-261)\tau^2\\ & +8(13m^2+80m+183)\rho^2-3(13m^2+80m-117)R^2)+O(r^{m+3})\}.\end{aligned}$$

Theorem 2.7: *Let* $(D,B)=(\tilde\Delta,B_D)$.

$$\begin{aligned}(a)\quad & a_2(\Delta,B_D,r)=\tfrac{c_2(m)}{(m-1)}\{v(m)r^{m-2}2(m-1)^2-r^m\alpha(m)(2m^2+3m-14)\tau\\ & +r^{m+2}\beta(m)(18(2m^2+9m-26)\Delta\tau+5(2m^2+10m-30)\tau^2\\ & +8(2m^2+4m-6)\rho^2-3(2m^2+4m-6)R^2+O(r^{m+4})\}.\\ (b)\quad & a_3(\tilde\Delta,B_D,r)=-\tfrac{c_3(m)}{(m-1)}\{v(m)r^{m-3}(m-1)^2(7m-17)\\ & -r^{m-1}\alpha(m)(7m^3+45m^2-199m+3)\tau\\ & +r^{m+1}\beta(m)(18(7m^3+105m^2-151m-681)\Delta\tau\\ & +5(7m^3+121m^2-191m-801)\tau^2\\ & +8(7m^2+32m+21)(m-1)\rho^2\\ & -3(7m^2+32m+81)(m-1)R^2)+O(r^{m+3})\}\ .\end{aligned}$$

Theorem 2.8: *Let* $(D,B) = (\tilde{\Delta}, B_{Ro})$.

(a) $a_2(\tilde{\Delta}, B_{Ro}, r) = \frac{c_2(m)}{m-1}\{v(m)r^{m-2}(-4m^2+14m-10)$
$-r^m\alpha(m)(-4m^2+9m-2)\tau$
$+r^{m+2}\beta(m)(18(-4m^2+3m+10)\Delta\tau + 5(-4m^2+4m+6)\tau^2$
$+8(-4m^2-2m+30)\rho^2 - 3(-4m^2-2m+30)R^2 + O(r^{m+4})\}$.

(b) $a_3(\tilde{\Delta}, B_{Ro}, r) = -\frac{c_3(m)(m)}{(m-1)^2}\{r^{m-3}v(m)(m-1)(13m^3-85m^2+179m-107)$
$+r^{m-1}\alpha(m)(m-1)(-13m^3-15m^2+253m-273)$
$+r^{m+1}\beta(m)\{18(13m^4+86m^3-496m^2+346m+51)\,\Delta\tau$
$+5(13m^4+102m^3-504m^2-214m+987)\tau^2$
$+8(13m^4+6m^3-156m^2+146m+471)\rho^2$
$-3(13m^4+6m^3-456m^2+1226m-789)R^2\} \;+\; O(r^{m+3})\}$.

We can use these calculations to draw some conclusions in spectral geometry. Let M_i be Riemannian manifolds for $i = 1,2$. We subscript to distinguish invariants; thus τ_i denotes the scalar curvature function on M_i for example. Let M_1 be a homogeneous manifold. Then τ_1, ρ_1^2, and R_1^2 are constant so $\Delta\tau_1 = 0$. We note spec$(D,B)_{B(x_1,r)}$ does not depend on the choice of x_1.

Theorem 2.9: *Let* $(D,B) \in \{(\Delta, B_D), (\Delta, B_N), (\tilde{\Delta}, B_D), (\tilde{\Delta}, B_{Ro})\}$. *Suppose*

$$spec(D,B)_{B(x_1,r)} = spec(D,B)_{B(x_2,r)}$$

for all sufficiently small r *and for all* $x_1 \in M_1$ *and* $x_2 \in M_2$. *Then* $\tau_1 = \tau_2$, $\rho_1^2 = \rho_2^2$, *and* $R_1^2 = R_2^2$.

Proof: We equate the asymptotic expansion of a_0 given by Theorem 2.4 for M_1 and M_2. Comparing the coefficients of r^{m+2} shows $\tau_1(x_1) = \tau_2(x_2)$ for all $x_1 \in M_1$ and $x_2 \in M_2$. Hence, τ_1 and τ_2 are constant and $\Delta\tau_1 = \Delta\tau_2 = 0$. Comparing the coefficient of r^{m+4} then yields

$$8\rho_1^2 - 3R_1^2 = (8\rho_2^2 - 3R_2^2)(x_2) \tag{2.16}$$

for all $x_2 \in M_2$. The asymptotic expansions of a_1 and a_2 provide no additional information. We use the asymptotic expansion of a_3 to see

$$\epsilon_1(8\rho_1^2 - 3R_1^2) + \epsilon_2 R_1^2 = \{\epsilon_1(8\rho_2^2 - 3R_2^2) + \epsilon_2 R_2^2\}(x_2) \tag{2.17}$$

where by inspection $\epsilon_2 \neq 0$. We combine (2.16) and (2.17) to complete the proof. ■

The Riemannian geometry generally is not determined by the linear and quadratic curvature invariants. In [FV], for example, it is shown CP^8 and the Cayley plane have the same τ, ρ^2, and R^2. Similar examples can be found in [L], [KTV] and [TV].

Since the linear and quadratic invariants are spectrally determined, the following Corollary is immediate.

Corollary 2.10: *Let M_i be as in Theorem 2.9.*
(a) If M_1 is Einstein, then M_2 is Einstein.
(b) If M_1 is conformally flat, then M_2 is conformally flat.
(c) If M_1 has constant sectional curvature c, then M_2 has constant sectional curvature c.

In a similar way as in [CVH] we have characterizations of the other rank 1 symmetric spaces.

Theorem 2.11: *Let M_i be as in Theorem 2.9.*
(a) If M_i are Kähler manifolds, and if M_1 is with constant holomorphic sectional curvature, then M_2 has the same constant holomorphic sectional curvature.
(b) Let the holonomy group of M_2 be a subgroup of $Sp(n) \cdot Sp(1)$ and let M_1 be $\mathbf{QP}^n(\nu)$ or its noncompact dual. Then M_2 is locally isometric to $\mathbf{QP}^n(\nu)$ or its noncompact dual.

It is not necessary to formulate a similar theorem for manifolds with holonomy group contained in $Spin(9)$, because such manifolds are automatically flat or locally isometric to the Cayley plane or its noncompact dual (see [A] and [BGr]).

References

[A] D. V. Alekseevskij, "On holonomy groups of Riemannian manifolds," Ukrain. Math. Z. **19** (1967), 100-104.

[ABC] P. Amsterdamski, A. Berkin, and D. O'Connor, "b_8 'Hamidew' coefficient for a scalar field", Class. Quantum Grav. **6** (1989), 1981-1991.

[Av] I. Avramidi, "The covariant technique for calculation of one-loop effective action," to appear in Nucl. Phys. B.

[BG] T. Branson and P. Gilkey, "The asymptotics of the Laplacian on a manifold with boundary," *Communications on PDE,* **15** (1990), 245-272.

[BO] T. Branson and B. Orsted, "Conformal indices of Riemannian manifolds," *Compositio Math.* **60**(1986), 261-293.

[BGr] R. B. Brown and A. Gray, "Manifolds whose holonomy group is subgroup of *Spin*(9)," Differential geometry (in honor of K. Yano), Tokyo, 1972, 41-59.

[CVH] B.Y. Chen and L. Vanhecke, "Differential geometry of geodesic spheres," *Journal für die reine und angewandte Mathematik,* **325**(1981), 28-67.

[FV] M. Ferraratti and L. Vanhecke, "Curvature invariants and symmetric spaces, " in preparation.

[G] P. Gilkey, "The spectral geometry of a Riemannian manifold," *J. Diff. Geom.* **10**(1975), 601-618.

[Gr] A. Gray, "The volume of a small geodesic ball in a Riemannian manifold," *Michigan Math. J.* **20**(1973), 329-344.

[GV] A. Gray and L. Vanhecke, "Riemannian geometry as determined by the volumes of small geodesic balls," *Acta Math.* **142**(1979), 157-198.

[KP] L. Karp and M. Pinsky, "The first eigenvalue of a small geodesic ball in a Riemannian manifold," *Bull. Sc. Math* **111** (1987), 229-239.

[KCD] G. Kennedy, R. Critchley, and J.S.Dowker, "Finite temperature field theory with boundaries: stress tensor and surface action renormalization," *Annals of Physics* **125**(1980), 346-400.

[KTV] O. Kowalski, F. Tricerri, and L. Vanhecke, "Curvature homogenous Riemannian manifolds," *J. Math. Pures Appl.*, to appear.

[L] F. Lastaria, "Homogeneous metrics with the same curvature," *Simon Stevin,* to appear.

[Le] A. Levy-Bruhl, "Invariants infinitesimaux", *C. R. Acad. Sc. Paris. T* **279** (1974), 197-200.

[MD] I. Moss and J.S. Dowker, "The correct B_4coefficient," (preprint).

[MS] H. McKean and I. Singer, "Curvature and the eigenvalues of the Laplacian," *J. Diff. Geom.* **1**(1967), 43-69.

[Sa] T. Sakai, "On eigenvalues of Laplacian and curvature of Riemannian manifolds," *Tohoku Math J* **23**(1971), 589-603.

[TV] F. Tricerri and L. Vanhecke, "Curvature homogenous Riemannian spaces," *Ann. Sc. Ecole Norm. Sup.* **22**(1989), 535-554.

Novica Blažić
Faculty of Mathematics
University of Belgrade
Studentski trg 16
P.B. 550
11000 Belgrade
Yugoslavia

Neda Bokan
Faculty of Mathematics
University of Belgrade
Studentski trg 16
P.B. 550
11000 Belgrade
Yugoslavia

Peter Gilkey
Mathematics Dept.
University of Oregon
Eugene Oregon 97403
USA

MINIMAL IMMERSIONS OF $\mathbb{R}P^2$ INTO $\mathbb{C}P^n$

by J.Bolton, W.M. Oxbury, L.Vrancken and L.M. Woodward

In this paper(†) we consider minimal immersions of the real projective plane $\mathbb{R}P^2$ into the complex projective space $\mathbb{C}P^n$ of constant holomorphic sectional curvature 4. Since $\mathbb{R}P^n$ with its standard metric of constant curvature 1 sits in $\mathbb{C}P^n$ as a totally geodesic submanifold this includes the case of minimal immersions of $\mathbb{R}P^2$ into $\mathbb{R}P^n$. It is clear that all the immersions under consideration can be studied in the context of minimal immersions of S^2 into S^n (with its standard metric) and $\mathbb{C}P^n$ with certain symmetry properties and this is what we shall do. For questions concerning the liftings of minimal immersions of $\mathbb{R}P^2$ into $\mathbb{R}P^n$ to minimal immersions of S^2 into S^n we refer the reader to [11].

The motivation for considering such maps is two-fold:

(a) This class of immersions contains the most beautiful and most studied immersions of S^2 into S^n, namely the Veronese immersions, which are characterised by having induced metrics of constant curvature. Indeed our initial motivation was to try to prove the Simon conjecture [15] (which seeks to characterise the Veronese immersions by a pinching condition on the curvature of the induced metric) for this special class of immersions.

(b) An important aspect in the study of minimal immersions (and more generally harmonic maps) of S^2 into S^n is the role of the higher order singularities in the sense of Chern [8]. (See also [4]). These are rather complicated in general but for the class of immersions considered here the singularity pattern determines the immersion up to ambient isometries almost completely. There is good reason to hope that the approach initiated here will lead to a much better understanding of the space of all minimal immersions (or more generally harmonic maps) of S^2 into S^n. The case $n = 4$ has been studied by Loo [14] and Verdier [16].

The approach we adopt is that initiated by Calabi [7] for minimal immersions of S^2 into S^n, and extended by others [9,10,12] to the case of minimal immersions of S^2 into $\mathbb{C}P^n$, of relating the given immersion to a corresponding holomorphic curve called its directrix curve. The outline

---

(†) This consists of two separate pieces of joint work, one by the first, second and fourth authors and the other by the first, third and fourth.

of the paper is as follows. In §1 we consider some basic facts about holomorphic curves in $\mathbb{C}P^n$ and their higher order singularity types including two key theorems. Then in §2 we recall the relationship between minimal immersions of S^2 into $\mathbb{C}P^n$ and holomorphic curves. In §3 we then deal with minimal immersions of $\mathbb{R}P^2$ into $\mathbb{C}P^n$ and state the main results. These are strong congruence results and in certain cases give explicit formulae for such immersions.

Readers unfamiliar with holomorphic curves may find [13] a useful reference for this material.

§1 Holomorphic curves

Let $\psi : S^2 \to \mathbb{C}P^n$ be a linearly full holomorphic curve, where S^2 denotes the Riemann sphere. (We recall that a map $\psi : S^2 \to \mathbb{C}P^n$ is linearly full if there is no proper linear subspace V of $\mathbb{C}^{n+1}$ such that $\psi(S^2) \subseteq P(V)$). We will say that ψ_0 has a <u>higher order singularity</u> of type $(u_1,\ldots,u_n)$ at $x \in S^2$ if for each $p = 1,\ldots,n$ the $(p-1)^{st}$ osculating curve [13] of ψ_0 has a singularity of index u_p at x. In terms of a local complex coordinate z on S^2 we may write $\psi(z) = [f(z)]$, where f(z) is a non-vanishing $\mathbb{C}^{n+1}$-valued polynomial function and $[f(z)]$ denotes the line determined by f(z). If $f^{(p)}$ denotes the p-th derivative of f with respect to z then $f,f^{(1)},\ldots,f^{(n)}$ are linearly independent at all but a finite set of points, namely the set $Z(\psi)$ of points where the higher order singularities of ψ_0 occur. If the set $Z(\psi)$, possibly including ∞, has cardinality k we say that ψ is <u>k-point ramified</u>. Assume ψ_0 has a higher order singularity at $z = a$ of type $(u_1,\ldots u_n)$. If, for $p = 1,\ldots,n$, we put $r_p = pu_1 + (p-1)u_2 + \ldots + u_p$ then

$$f\wedge f^{(1)}\wedge\ldots\wedge f^{(p)}(z) = (z-a)^{r_p}\, w_p(z)$$

for some $w_p(z)$ with $w_p(a) \neq 0$. In particular, $Z(\psi)$ is a finite set and, if $Z(\psi) = \{a_1,\ldots,a_k\}$, we call the set

$$\{(a_j\ ;\ u_1(a_j),\ldots,u_n(a_j))\}_{j=1,\ldots,k}$$

the <u>singularity type</u> of ψ. Note that projectively equivalent holomorphic curves have the same singularity type. The converse is false but as we shall see below a holomorphic curve is determined (modulo projective equivalence) up to a finite number of possibilities by its singularity type.

First, however, in order to elucidate the ideas so far and in order to explain the first key theorem we consider the case of holomorphic curves which are k-point ramified for $k \leq 2$. Particular examples of these are given by holomorphic curves which have <u>S^1-symmetry</u>; that is to say the group of isometries of the induced metric on S^2 contains a circle subgroup. In this case the higher order singularities, if any, are at the fixed points of the S^1-action.

<u>Theorem 1</u> [3] *Let $\psi : S^2 \to \mathbb{C}P^n$ be a linearly full holomorphic curve which is* k-point *ramified for* $k \leq 2$. *Then* $k = 0$ *or* 2 *and, assuming the singularities if any are at* $z = 0$ *and* $z = \infty$, ψ *is given by*

$$\psi(z) = \Big[\sum_{p=0}^{n} z^{k_1+\ldots+k_p}\, \underline{b}_p\Big]$$

where $\underline{b}_0,\ldots,\underline{b}_n$ *is a basis for* $\mathbb{C}^{n+1}$, $k_1,\ldots,k_n$ *are strictly positive integers and* ψ_0 *has singularity type* $\{(0;\ k_1-1,\ldots,k_n-1),$ $(\infty;\ k_n-1,\ldots,k_1-1)\}$. *Moreover* ψ *has* S^1*-symmetry (with fixed points of the* S^1*-action at* $z = 0$ *and* $z = \infty$*) if and only if* $\underline{b}_0,\ldots,\underline{b}_n$ *are mutually orthogonal.*

Now we turn to the other main theorem of this section.
<u>Theorem 2</u> *Let $\psi : S^2 \to \mathbb{C}P^n$ be a (linearly full) holomorphic curve. Then the set of holomorphic curves, modulo projective equivalence, having the same singularity type as ψ is finite.*

The proof which is lengthy but elementary depends on a series of steps which we now briefly sketch. (Care should be taken not to confuse the notation used here with that used in later sections).

(i) For a fixed integer $d \geq n$ let G be the set of linearly full holomorphic curves $\varphi : S^2 \longrightarrow \mathbb{C}P^n$ of degree d, modulo projective equivalence. Then G may be naturally identified with a Zariski-open set of the Grassmannian $\mathrm{Grass}_{n+1}\, H^0(\mathcal{O}(d))$ of (n+1)-planes in the (d+1)-dimensional vector space $H^0(\mathcal{O}(d))$ of holomorphic sections of the line bundle $\mathcal{O}(d)$ of degree d over S^2, the inclusion $G \longrightarrow \mathrm{Grass}_{n+1}\, H^0(\mathcal{O}(d))$ being given by

$$< \varphi > \in G \longmapsto \operatorname{span}\{\varphi_0,\ldots,\varphi_n\} \in \operatorname{Grass}_{n+1}H^0(\mathcal{O}(d))$$

where $<\varphi>$ is the class of $\varphi = [\varphi_0,\ldots,\varphi_n]$.
In fact the image of G in $\operatorname{Grass}_{n+1} H^0(\mathcal{O}(d))$ is the complement of the set of subspaces $V \subset H^0(\mathcal{O}(d))$ such that for some $a \in S^2$, $\varphi(a) = 0$ for every $\varphi \in V$. Such subspaces correspond to maps $\varphi : S^2 \to \mathbb{C}P^n$ of degree less than d.
If $\Lambda^{n+1} = \Lambda^{n+1} H^0(\mathcal{O}(d))$ then G is embedded in $\mathbb{P}\Lambda^{n+1}$ via the Plücker embedding defined by $<\varphi> \longmapsto [\varphi_0\wedge\ldots\wedge\varphi_n]$.

(ii) For each point $w \in \mathbb{C}$ and each integer $p \geq 0$ there is a hyperplane $\sigma^{(p)}(w)$ in $\mathbb{P}\Lambda^{n+1}$ determined as follows. If $e_0,\ldots,e_d$ is a basis for $H^0(\mathcal{O}(d))$ then the elements

$$e_{(i_0,\ldots,i_n)} = e_{i_0}\wedge\ldots\wedge e_{i_n}, \quad 0 \leq i_0 <\ldots< i_n \leq d,$$

form a basis for Λ^{n+1}. Define

$$\Phi_w^{(p)} : \Lambda^{n+1} \longrightarrow \mathbb{C} \text{ by}$$

$$\Phi_w^{(p)}\Big[\sum_I b_I e_I\Big] = \sum_I b_I \frac{d^p}{dz^p} E_I(w)$$

where the sum is over the multi-indices $I = (i_0,\ldots,i_n)$ with $0 \leq i_0 <\ldots < i_n \leq d$, and

$$E_I = \det \begin{bmatrix} e_{i_0} & \ldots\ldots & e_{i_n} \\ \vdots & & \vdots \\ e_{i_0}^{(n)} & \ldots\ldots & e_{i_n}^{(n)} \end{bmatrix}$$

(so that $E_I \in H^0(\mathcal{O}(N))$, where $N = (n+1)(d-n)$).
The map $\Phi_w^{(p)}$ is clearly linear and $\sigma^{(p)}(w)$ is the hyperplane in $\mathbb{P}\Lambda^{n+1}$ determined by $\ker \Phi_w^{(p)}$.

(iii) Define $\beta : G \longrightarrow \mathbb{P}H^0(\mathcal{O}(N))$ by

$$\beta <\varphi> = \left[\det \begin{bmatrix} \varphi_0 & \ldots & \varphi_n \\ \vdots & & \vdots \\ \varphi_0^{(n)} & \ldots & \varphi_n^{(n)} \end{bmatrix}\right].$$

If φ has singularity type $\{(z_i;u_{1,i},\ldots,u_{n,i})\}_{i=1,\ldots,k}$, define

$$s_i = \sum_{p=1}^{n} (n-p+1)u_{p,i} \quad , \ i = 1,\ldots,k.$$

Then $\beta < \varphi > = [h(z)]$, where

$$h(z) = (z-z_1)^{s_1}\ldots(z-z_k)^{s_k}.$$

(We note here that by the Plücker formulae $s_1+\ldots+s_k =$ $(n+1)(d-n) = N$ so that $h(z)$ determines an element of $H^0(\mathcal{O}(N))$).

(iv) The theorem is now proved by observing that

$$\beta^{-1}(h(z)) = G \cap \Sigma ,$$

where Σ is the intersection of the N hyperplanes

$$\sigma^{(j)}(z_i) \ , \ 0 \leq j \leq s_i-1 \ , \ i =1,\ldots,k,$$

and consequently that the number of points in $\beta^{-1}(h(z))$ is less than or equal to the degree of the Plücker embedding $G \longrightarrow \mathbb{P}\Lambda^{n+1}$. It then follows that the number of elements in G having the same singularity type as ψ is finite.

Remark: One can show that equality holds over polynomials $h(z)$ with only simple zeros. One can interpret this by saying that one has a branched cover

$$\beta : G \longrightarrow \mathbb{P}H^0(\mathcal{O}(N))$$

branched over the discriminant locus. In fact a refinement of the above arguments can be used to analyse the branching behaviour more closely. This then enables one to count the precise number of points in any fibre of β.

§2 Minimal immersions of S^2

Let $\psi_0 : S^2 \longrightarrow \mathbb{C}P^n$ be a linearly full holomorphic curve and, as in §1, write $\psi_0(z) = [f_0(z)]$ for some non-vanishing polynomial function $f_0(z)$. The Frenet frame of ψ_0 is the sequence $\psi_0,\ldots,\psi_n : S^2 \longrightarrow \mathbb{C}P^n$ of functions defined by $\psi_p(z) = [f_p(z)]$ where $f_0,\ldots,f_n$ are obtained from $f_0,\ldots,f_0^{(n)}$ by Gram-Schmidt orthogonalisation (It is clear that some care is needed here: see [6,10,17] for details).

We recall the following:

Fundamental Theorem [9,10,12]. *Let $\psi_0 : S^2 \longrightarrow \mathbb{C}P^n$ be a linearly full holomorphic curve. Then each element $\psi_p : S^2 \longrightarrow \mathbb{C}P^n$ of the Frenet frame $\psi_0,\ldots,\psi_n$ of ψ_0 is a (branched) conformal minimal immersion. Conversely given a linearly full (branched) conformal minimal immersion $\psi : S^2 \longrightarrow \mathbb{C}P^n$*

ψ) *and a unique integer* p, $0 < p < n$, *such that* ψ *is the* p-*th element* ψ_p *of the Frenet frame of* ψ_0.

Remark: Recall that any immersion of S^2 can be made conformal by first applying a suitable diffeomorphism of S^2 to itself.

If $\varphi : S^2 \longrightarrow S^n$ is an immersion then so is $\psi = i\,\pi\,\varphi : S^2 \longrightarrow \mathbb{C}P^n$, where $\pi : S^n \longrightarrow \mathbb{R}P^n$ is the standard double cover and $i : \mathbb{R}P^n \longrightarrow \mathbb{C}P^n$ is the standard inclusion. Moreover φ is minimal if and only if $\psi = i\,\pi\,\varphi$ is minimal since i is totally geodesic. We now characterise those minimal immersions ψ which arise in this way. To do this we note that if $\psi_0,\ldots,\psi_n$ is the Frenet frame of a holomorphic curve then ψ_n is antiholomorphic, and it is not hard to show that $\overline{\psi}_n,\ldots,\overline{\psi}_0$ is the Frenet frame of $\overline{\psi}_n$. This leads to the following theorem.

Theorem 3 *Let* $\psi : S^2 \longrightarrow \mathbb{C}P^n$ *be a linearly full minimal immersion and let* $\psi_0,\ldots,\psi_n$ *be the Frenet frame of its directrix curve* ψ_0. *Then* $\psi = i\,\pi\,\varphi$ *for some* $\varphi : S^2 \longrightarrow S^n$ *if and only if*

(i) $n = 2m$,

(ii) $\psi = \psi_m$,

(iii) $\psi_0 = \overline{\psi}_n$.

Remark: There are several different formulations of this result (see [1,10]) but the one given here is the most convenient for our purposes.

In the case of (linearly full) minimal immersions $\varphi : S^2 \longrightarrow S^n$ the higher order invariants in the sense of Chern [8] correspond precisely to the higher order invariants of $\psi = i\pi\varphi$ discussed in §1. The interested reader will find details in [4]. It is perhaps worth remarking that all these different equivalent conditions are extremely hard to check for particular holomorphic curves.

§3 Minimal immersions of $\mathbb{R}P^2$ into $\mathbb{C}P^n$

Given an immersion of $\mathbb{R}P^2$ into $\mathbb{C}P^n$ we obtain in a natural way an immersion $\psi : S^2 \longrightarrow \mathbb{C}P^n$. It is not hard to show (c.f.[3]) that without loss of generality we may assume that ψ is conformal.

A complex coordinate on S^2 obtained by stereographic projection from a point $x_0 \in S^2$ onto the plane through the origin of $\mathbb{R}^3$ orthogonal to x_0 will be called a stereographic complex coordinate. We note that, with respect to such a complex coordinate z, the antipodal map is given by

$z \mapsto -1/\bar{z}$. Hence $\psi : S^2 \longrightarrow \mathbb{C}P^n$ factors through $\mathbb{R}P^2$ if and only if $\psi(z) = \psi(-1/\bar{z})$. The following results are proved in [3].

Theorem 4 *Let $\psi : S^2 \longrightarrow \mathbb{C}P^n$ be a linearly full minimal immersion and let $\psi_0, \ldots, \psi_n$ be the Frenet frame of its directrix curve ψ_0. Then ψ factors through $\mathbb{R}P^2$ if and only if*

(i) $n = 2m$,

(ii) $\psi = \psi_m$,

(iii) $\psi_0(z) = \psi_n(-1/\bar{z})$ *where z is a stereographic complex coordinate.*

Theorem 5 *Let $\psi : S^2 \longrightarrow \mathbb{C}P^n$, $n = 2m$, be a linearly full minimal immersion which factors through $\mathbb{R}P^2$, and suppose that the directrix curve ψ_0 is k-point ramified for $k < 2$. Let z be a stereographic complex coordinate such that if $k \neq 0$ then ψ_0 has a higher order singularity at $z = 0$. Then ψ_0 is given by*

$$\psi_0(z) = \left[\sum_{p=0}^{n} \sqrt{\lambda_p}\, z^{k_1 + \ldots + k_p}\, \underline{e}_p \right] \qquad (\dagger)$$

where $\underline{e}_0, \ldots, \underline{e}_n$ is a unitary basis for $\mathbb{C}^{n+1}$, $k_1, \ldots, k_n$ are odd positive integers, and

$$\lambda_p = \frac{\prod_{1<i<j<n} (k_i + \ldots + k_j)}{\prod_{i=1}^{p} (k_i + \ldots + k_p) \prod_{i=1}^{n-p} (k_{p+1} + \ldots + k_{n-i+1})}, \quad p=0,\ldots,n.$$

In particular ψ is S^1-symmetric and is uniquely determined, up to holomorphic isometries of $\mathbb{C}P^n$, by its singularity type. Furthermore, $\psi = i\,\pi\,\varphi$ for some $\varphi : S^2 \longrightarrow S^n$ if and only if ψ_0 can be written in the form $(\dagger)$ with $k_p = k_{n-p+1}$ for $p = 1, \ldots, n$ and $\underline{e}_p = (-1)^p\, \bar{\underline{e}}_{n-p}$ for $p = 0, \ldots, n$.

Remark: In the case when $k = 0$ so that $k_1 = \ldots = k_n = 1$ we have $\lambda_p = \binom{n}{p}$, and ψ_m is essentially the Veronese immersion with constant curvature $\dfrac{2}{m(m+1)}$. (For details see [3,5]).

The proof of Theorem 5 follows from Theorems 1,3 and 4. The only additional idea needed is that of determining ψ_n given ψ_0, which we now explain since it is also required for the proof of the next theorem.

Recall that the standard Hermitian inner product $<\ ,\ >$ on $\mathbb{C}^{n+1}$ defined by

$$< (x_0,\ldots,x_n)\ ,\ (y_0,\ldots,y_n) > = \sum_{p=0}^{n} x_p\bar{y}_p$$

extends to a Hermitian inner product on $\Lambda^r\mathbb{C}^{n+1}$ for $r = 0,\ldots,n$. Then the Hodge star operator $* : \Lambda^r\mathbb{C}^{n+1} \longrightarrow \Lambda^{n+1-r}\mathbb{C}^{n+1}$ is the sesquilinear map defined by $<\beta,*\alpha>\omega = \alpha\wedge\beta$, $\alpha \in \Lambda^r\mathbb{C}^{n+1}$, $\beta \in \Lambda^{n+1-r}\mathbb{C}^{n+1}$, where $\omega = e_0\wedge\ldots\wedge e_n$ with $e_0,\ldots,e_n$ the standard unitary basis of $\mathbb{C}^{n+1}$. It follows that if $u_0,\ldots,u_{n-1}$, v are linearly independent vectors in $\mathbb{C}^{n+1}$ with $<u_p,v> = 0$ for $p = 0,\ldots,n-1$ then $[v] = [*(u_0\wedge\ldots\wedge u_{n-1})]$ in $\mathbb{C}P^n$. Thus given a linearly full holomorphic curve $\psi_0 : S^2 \longrightarrow \mathbb{C}P^n$ then, writing $\psi_0 = [f_0]$, we have $\psi_n = [f_n] = [*(f_0\wedge\ldots\wedge f_0^{(n-1)})]$ since f_n is orthogonal to $f_0,\ldots,f_0^{(n-1)}$.

Moreover if $\tilde{\psi}_0$ is a holomorphic curve projectively equivalent to ψ_0, then $\tilde{\psi}_0 = [Af_0]$ for some $A \in GL(n+1;\mathbb{C})$. If the Frenet frame of $\tilde{\psi}_0$ is $\tilde{\psi}_0,\ldots,\tilde{\psi}_n$ with $\tilde{\psi}_p = [g_p]$, where $g_0 = Af_0$ and $g_0,\ldots,g_n$ are obtained from $g_0,\ldots,g_0^{(n)}$ by Gram-Schmidt orthogonalisation, then

$$\begin{aligned}[g_n] &= [*(Af_0\wedge\ldots\wedge Af_0^{(n-1)})]\\ &= [*\ \Lambda^n A *^{-1} *f_0\wedge\ldots\wedge f_0^{(n-1)})]\\ &= [(\mathrm{adj}\ A)^\dagger f_n].\end{aligned}$$

Thus,

$$\tilde{\psi}_n = [(\mathrm{adj}\ A)^\dagger]\psi_n.$$

We have thus proved:

<u>Lemma</u> *Let $\psi_0 : S^2 \longrightarrow \mathbb{C}P^n$ be a linearly full holomorphic curve and let $g \in PGL(n+1;\mathbb{C})$. If ψ_n denotes the last element of the Frenet frame of ψ_0 then the last element of the Frenet frame of $g\,\psi_0$ is $(\mathrm{adj}\ g)^\dagger\,\psi_n$.*

We may now prove the following theorem.

<u>Theorem 6</u>. *Let ψ, $\tilde{\psi} : S^2 \longrightarrow \mathbb{C}P^n$ be linearly full, minimal immersions which factor through $\mathbb{R}P^2$. Then $\psi,\tilde{\psi}$ are ambient isometric (by a holomorphic isometry of $\mathbb{C}P^n$) if and only if their directrix curves are projectively equivalent.*

Proof Suppose the directrix curves are projectively equivalent, say $\tilde{\psi}_0 = g\,\psi_0$. It follows from Theorem 4 and the lemma that

$$g\,\psi_0(z) = (\text{adj } g)^{\dagger}\,\psi_n(-1/\bar{z}) = (g^{-1})^{\dagger}\,\psi_0(z),$$

so that $g^{\dagger}g\,\psi_0 = \psi_0$. Since ψ_0 is linearly full $g \in PU(n+1)$. The converse is trivial.

Remark The condition of factoring through $\mathbb{R}P^2$ is crucial here. The result is not true in general. However, one may similarly prove that if $\varphi, \tilde{\varphi}: S^2 \longrightarrow S^n$, $n = 2m$, are linearly full minimal immersions then their directrix curves are projectively equivalent if and only if they differ by an element of $SO(n+1; \mathbb{C})$ (c.f.[1]). This, together with Theorem 2, is a first step in dealing with the space of minimal immersions of S^2 into S^n.

Finally, combining Theorems 2 and 6, we have the following partial generalisation of Theorem 5.

Theorem 7 *Let $\psi : S^2 \longrightarrow \mathbb{C}P^n$ be a linearly full minimal immersion which factors through $\mathbb{R}P^2$. Then up to holomorphic isometries of $\mathbb{C}P^n$ there are only finitely many minimal immersions of S^2 into $\mathbb{C}P^n$ which factor through $\mathbb{R}P^2$ and have the same singularity type as ψ.*

References

1. Barbosa, J.L. : On minimal immersions of S^2 into S^{2m}. Trans. Amer. Math.Soc. 210, 75-106 (1975).
2. Bolton, J., Jensen, G.R., Rigoli, M. and Woodward, L.M. : On conformal minimal immersions of S^2 into $\mathbb{C}P^n$. Math. Ann. 279, 599-620 (1988).
3. Bolton, J., Vrancken, L. and Woodward, L.M. : Minimal immersions of S^2 and $\mathbb{R}P^2$ into $\mathbb{C}P^n$ with few higher order singularities. To appear in Math. Proc. Camb. Phil. Soc.
4. Bolton, J. and Woodward, L.M. : On immersions of surfaces into space forms. Soochow J. of Mathematics 14, 11-31 (1988).
5. Bolton, J. and Woodward, L.M. : On the Simon conjecture for minimal immersions with S^1-symmetry. Math. Z. 200, 111-121 (1988).
6. Bolton, J. and Woodward, L.M. : Congruence theorems for harmonic maps from a Riemann surface into $\mathbb{C}P^n$ and S^n. To appear in J. Lond. Math. Soc.
7. Calabi, E. : Minimal immersions of surfaces into Euclidean spheres.

8. Chern, S.S. : On the minimal immersions of the two-sphere in a space of constant curvature. In: Problems in analysis pp.27-40, Princeton : Princeton University Press 1970.

9. Din, A.M. and Zakrzewski, W.J. " General classical solutions in the $\mathbb{C}P^{n-1}$ model. Nuclear Phys. B. 174, 397-406 (1980).

10. Eells, J. and Wood, J.C. : Harmonic maps from surfaces to complex projective spaces. Adv. Math. 49, 217-263 (1983).

11. Ejiri, N. : Equivariant minimal immersions of S^2 into $S^{2m}(1)$. Trans. Amer. Math. Soc. 297, 105-124 (1986).

12. Glaser, V. and Stora, R. : Regular solutions of the $\mathbb{C}P^n$ models and further generalisations. Preprint, CERN 1980.

13. Griffiths, P. and Harris, J. : Principles of algebraic geometry. London, New York: Wiley 1978.

14. Loo, B. : The space of harmonic maps of S^2 into S^4. Trans. Amer. Math. Soc. 313, 81-102 (1989).

15. Simon, U. : Eigenvalues of the Laplacian and minimal immersions into spheres. 115-120 in Differential Geometry, Cordero, L.A. (ed), Montreal: Pitman 1985.

16. Verdier, J.L. : Applications harmoniques de S^2 dans S^4, (preprint).

17. Wolfson, J.G. : Harmonic sequences and harmonic maps into complex Grassmann manifolds. J. Differ. Geom. 27, 161-178 (1988).

J. Bolton, W.M.Oxbury, L.M.Woodward
University of Durham
Department of Mathematical Sciences
Science Laboratories
South Road
Durham DH1 3LE
United Kingdom

L. Vrancken
Katholieke Universiteit Leuven
Faculteit Wetenschappen
Departement Wiskunde
Celestijnenlaan 200B
B-3030 Leuven
Belgium

Isoptics of a closed strictly convex curve

Waldemar Cieślak, Andrzej Miernowski, Witold Mozgawa

0. Introduction

The α-isoptic of a given convex curve consists of those points in the plane from which the curve is seen under the fixed angle α. Though isoptics alredy attracted the interest of the geometers a long time ago there remained several open questions concerning the general properties of these curves. The following considerations will give some results in that direction.

The regularity of isoptics will be discussed, including a simple construction for their tangents. A geometric condition for an isoptic in order to be convex will be established. Furthermore we present analytic condition for a closed curve in order to be an isoptic of a suitable convex curve. The general computations are applied to derive some integral formulas of Crofton type including an interesting proof of the well-known Crofton formula as well.

1. Remarks on smoothness of isoptics.

Let C be a plane, closed, strictly convex curve. We take a coordinate system with origin 0 in the interior of C. Let $p(t)$, $t \in [0, 2\pi]$, be the distance from 0 to the support line $l(t)$ to C perpendicular to the vector e^{it}. The function p is called a support function of the curve C. It is well known (cf. [2]) that the support function is differentiable and that the parametrization of C in terms of this function is given by

(1.1)
$$z(t) = p(t)e^{it} + \dot{p}(t)ie^{it} \quad \text{for} \quad t \in [0, 2\pi].$$

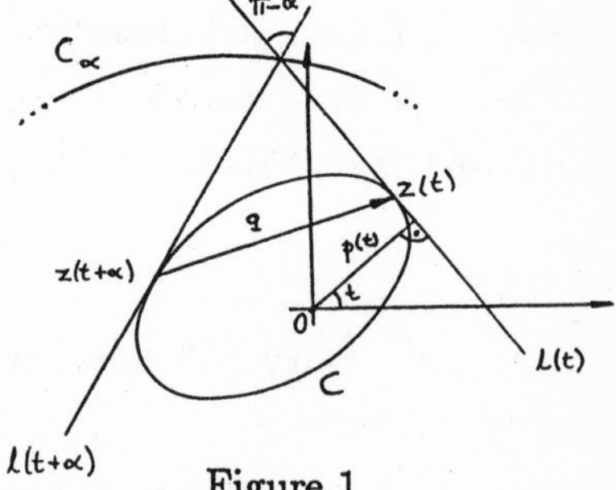

Figure 1

Let C_α be a locus of vertices of a fixed angle $\pi - \alpha$ formed by two support lines of the curve C. The curve C_α will be called an alpha-isoptic of C.

Next we introduce the following notations:

$$\begin{aligned} q(t) &= z(t) - z(t+\alpha), \\ b(t) &= [q(t), e^{it}], \\ B(t) &= [q(t), ie^{it}], \end{aligned} \tag{1.2}$$

where $[v, w] = ad - bc$ when $v = a + bi$ and $w = c + di$. With the above notations we obtain

$$b(t) = p(t+\alpha)\sin\alpha + \dot{p}(t+\alpha)\cos\alpha - \dot{p}(t) \tag{1.3a}$$

$$B(t) = p(t) - p(t+\alpha)\cos\alpha + \dot{p}(t+\alpha)\sin\alpha. \tag{1.3b}$$

It is convenient to parametrize the α-isoptic C_α by the same angle t

$$z_\alpha(t) = z(t) + \lambda(t)ie^{it} = z(t+\alpha) + \mu(t)ie^{i(t+\alpha)} \tag{1.4}$$

where

$$\lambda(t) = b(t) - B(t)\cot\alpha \tag{1.5a}$$

$$\mu(t) = -\frac{B(t)}{\sin\alpha}. \tag{1.5b}$$

Thus the equation of C_α has the form

$$z(t) = p(t)e^{it} + \{-p(t)\cot\alpha + \frac{1}{\sin\alpha}p(t+\alpha)\}ie^{it}.$$

Since the function involved are at least of class C^1 we observe that the given parametrization of an isoptic of a closed, strictly convex curve is of class C^1. This does not imply that the isoptic must be regular. In section 5 it will be proved that all α-isoptics of an oval (i.e. C^2 curve with nonvanishing curvature) are curves of class C^2.

2. Sine theorem for isoptics.

The sine theorem for ovals presented in [1] will be extended now to the class of all closed, strictly convex curves. Let us consider a tanget line to the isoptic C_α at the point $z_\alpha(t)$. Let ξ and η denote the angles indicated on the Fig. 2. We note that

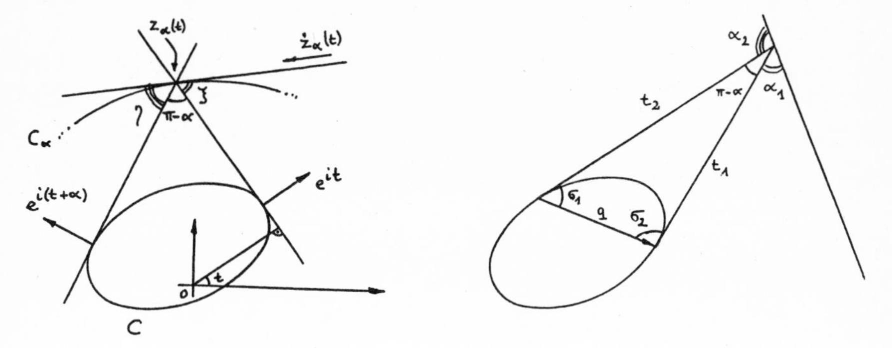

Figure 2 Figure 3

$$\sin\xi = -\frac{[\dot z_\alpha(t), ie^{it}]}{|\dot z_\alpha(t)|}. \tag{2.1}$$

Since

$$|\dot z_\alpha(t)| = \frac{|q(t)|}{\sin\alpha} \tag{2.2}$$

we have

$$\frac{|q|}{\sin\alpha} = \frac{\lambda}{\sin\xi}. \tag{2.3}$$

Similarly, we obtain

$$\sin\eta = \frac{[\dot{z}_\alpha(t), ie^{i(t+\alpha)}]}{|\dot{z}_\alpha(t)|} = -\frac{\mu \sin\alpha}{|q|}. \tag{2.4}$$

Finally, we have the following

Theorem 2.1. The following relations hold

$$\frac{|q|}{\sin\alpha} = \frac{|z_\alpha(t) - z(t)|}{\sin\xi} = \frac{|z_\alpha(t) - z(t+\alpha)|}{\sin\eta}. \tag{2.5}$$

We shall now describe a geometric procedure for a construction of a tangent line to an α-isoptic. With the notations of the Fig. 3. we have

$$\frac{|q|}{\sin\alpha} = \frac{t_1}{\sin\alpha_1} = \frac{t_2}{\sin\alpha_2}. \tag{2.6}$$

Using the classical sine theorem we obtain

$$\alpha_1 = \sigma_1 \quad \text{and} \quad \alpha_2 = \sigma_2. \tag{2.7}$$

Thus the construction is clear from the Fig.4.

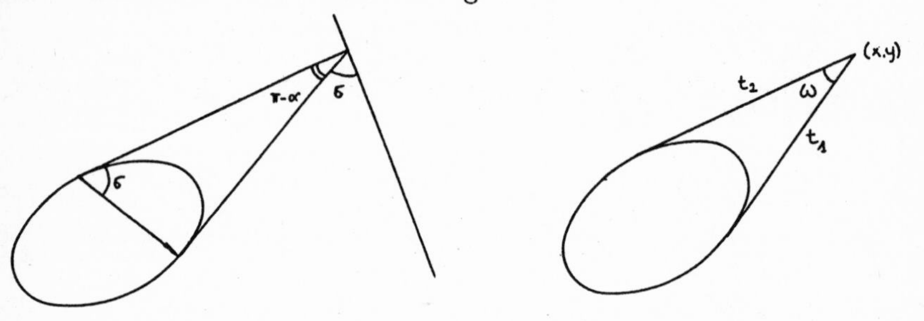

Figure 4

Figure 5

3. Crofton-type formulas and isoptics.

Let $\mathcal{L}$ be the exterior of a closed, strictly convex curve C. We define a mapping $F : (0,\pi) \times (0, 2\pi) \longrightarrow \mathcal{L} \setminus \{certain\ support\ half-line\}$ by the formula

$$F(\alpha, t) = z_\alpha(t). \tag{3.1}$$

The partial derivatives of F at (α, t) are given by

$$\frac{\partial F}{\partial \alpha} = -\frac{\mu(\alpha,t)}{\sin\alpha} ie^{it}, \tag{3.2a}$$

$$\frac{\partial F}{\partial t} = \lambda(\alpha,t)e^{it} + \{p(t) - \dot{p}(t)\cot\alpha + \frac{\dot{p}(t+\alpha)}{\sin\alpha}\}ie^{it} \tag{3.2b}$$

hence the jacobian determinant $F'(\alpha, t)$ of F at (α, t) is equal to

$$F'(\alpha,t) = -\frac{\lambda(\alpha,t)\mu(\alpha,t)}{\sin\alpha} > 0. \tag{3.3}$$

Using the formula for change of variables in multiple integrals, we obtain the well-known Crofton formula (see [3])

$$\int\int_{\mathcal{L}} \frac{\sin\omega}{t_1 t_2} dx dy = \int_0^{2\pi}\int_0^{\pi} d\alpha dt = 2\pi^2 \tag{3.4}$$

where notations are given by the Fig. 5. Similarly, under the same notations we obtain new Crofton-type formulas, namely

$$\int\int_{\mathcal{L}} \frac{\sin^2\omega}{t_1} dx dy = \int_0^{2\pi}\int_0^{\pi} B(\alpha,t) d\alpha dt = \pi L \tag{3.5}$$

and

$$\int\int_{\mathcal{L}} \frac{\sin^2\omega}{t_2} dx dy = \int_0^{2\pi}\int_0^{\pi} (b(\alpha,t)\sin\alpha + B(\alpha,t)\cos\alpha) d\alpha dt = \pi L \tag{3.6}$$

where L is the perimeter of C.

It should be noticed that other formulas that can be found in the literature can be easily proved by means of the diffeomorphism F. The results of this section extend the considerations of the paper [1].

4. Steiner centroid of an α-isoptic.

In this section we prove that the Steiner centroid of a given closed, strictly convex curve and its α-isoptic coincide.

Theorem 4.1. Let $t \longrightarrow z(t)$ be a closed strictly convex curve parametrized as in (1.1) and let $t \longrightarrow z_\alpha(t)$ be its α-isoptic. Then

$$\int_0^{2\pi} z(t)dt = \int_0^{2\pi} z_\alpha(t)dt. \tag{4.1}$$

Proof. By simple calculations we obtain

$$\int_0^{2\pi} z(t)dt = \int_0^{2\pi} p(t)e^{it}dt + \int_0^{2\pi} \dot{p}(t)ie^{it}dt = 2\int_0^{2\pi} p(t)e^{it}dt, \tag{4.2}$$

Similarly

$$\int_0^{2\pi} z_\alpha(t)dt = \int_0^{2\pi} p(t)e^{it}dt + \frac{i}{\sin\alpha}\left(\int_0^{2\pi} p(t+\alpha)ie^{it}dt - \cos\alpha\int_0^{2\pi} p(t)e^{it}dt\right)$$

$$= 2\int_0^{2\pi} p(t)e^{it}dt. \tag{4.3}$$

5. Convexity of isoptics. In this section we will show that all α-isoptics of an oval (i.e. a closed convex C^2 curve with nonvanishing curvature) are curves of class C^2. Moreover, in terms of the vector field q along the curve C we give a necessary and sufficient condition for an α-isoptic to be convex. Let C be an oval, then C is a closed, strictly convex curve. To determine the conditions for the convexity of isoptics we derive some identities involving the functions b and B. It is easy to verify that

$$\dot{b} = B + (p(t+\alpha) + \ddot{p}(t+\alpha))\cos\alpha - (p(t) + \ddot{p}(t)) \tag{5.1a}$$
$$\dot{B} = -b + (p(t+\alpha) + \ddot{p}(t+\alpha))\sin\alpha. \tag{5.1b}$$

We note that $R(t) = p(t) + \ddot{p}(t)$ is the radius of curvature of C at $z(t)$. Thus the system (5.1) can be rewritten in the form

$$\dot{b} = B + R(t+\alpha)\cos\alpha - R(t) \tag{5.2a}$$
$$\dot{B} = -b + R(t+\alpha)\sin\alpha. \tag{5.2b}$$

We recall that the equation of an α-isoptic C_α of the oval C is given by

$$z_\alpha(t) = z(t) + \lambda(t)ie^{it} \quad for \quad t \in [0, 2\pi], \tag{5.3}$$

where

$$\lambda = b - B\cot\alpha. \tag{5.4}$$

We shall now consider a tangent vector to C_α at $z_\alpha(t)$. By straightforward calculations we obtain

$$\dot{z}_\alpha(t) = -\lambda(t)e^{it} + \rho(t)ie^{it}, \tag{5.5}$$

where

$$\rho = B + b\cot\alpha. \tag{5.6}$$

Thus we obtain the following conclusion:
Theorem 5.1. An α-isoptic of an oval is a curve of class C^2.

The curvature k_α of the α-isoptic C_α can be computed from the formula (5.5), namely we have

$$\begin{aligned} k_\alpha(t) &= \frac{1}{|\dot{z}_\alpha(t)|^3}[\dot{z}_\alpha(t), \ddot{z}_\alpha(t)] \\ &= \frac{1}{(\rho(t)^2 + \lambda(t)^2)^{\frac{3}{2}}}\{2\rho(t)^2 + 2\lambda(t)^2 - R(t)\rho(t) + \lambda(t)R(t)\cot\alpha - \lambda(t)\frac{R(t+\alpha)}{\sin\alpha}\} \\ &= \frac{\sin\alpha}{|q(t)|^3}(2|q(t)|^2 - [q(t), \dot{q}(t)]). \end{aligned} \tag{5.7}$$

Hence we obtain the condition for an α-isoptic to be convex.
Theorem 5.2. An α-isoptic C_α of an oval C is convex if and only if

$$[q, \dot{q}] \le 2|q|^2. \tag{5.8}$$

Remark. The condition (5.8) is equivalent to the another one

$$|\frac{d}{dt}(\frac{q}{|q|})| \leq 2. \tag{5.9}$$

6. All α-isoptics are star-shaped.

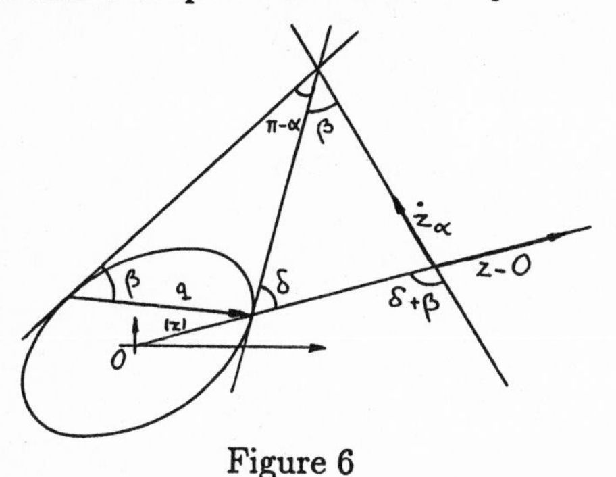

Figure 6

We put the Steiner centroid of C at the origin 0 of the coordinate system. Within the context of the previous section we have the following Fig. 6. Then the α-isoptic is star shaped if and only if

$$[z_\alpha - 0, \dot{z}_\alpha] \geq 0. \tag{6.1}$$

Thus

$$[z_\alpha, \dot{z}_\alpha] = [z, \dot{z}_\alpha] + \lambda^2 \tag{6.2}$$

and since $\delta + \beta$ is the angle that $z - 0$ forms with $\dot{z}_\alpha$ and $\delta + \beta \in [0, \pi]$, then $[z, \dot{z}_\alpha] \geq 0$ for each t and α. Now we can formulate the following theorem.

Theorem 6.1. An α-isoptic of an oval is star-shaped.

7. Conditions for a curve to be an α-isoptic.

W. Wunderlich in [5], [6] has considered curves with an isoptic circle and in [7] curves with an isoptic ellipse. In this section we give a necessary and sufficient condition for a curve to be an α-isoptic of a given oval. Our method differs essentially from the considerations of W. Wunderlich. Let W be a Jordan curve of class C^2, $t \longrightarrow w(t)$ for $t \in [0, 2\pi]$. We assume that the Steiner centroid of W lies at 0, that is

$$\int_0^{2\pi} w(t)dt = 0. \tag{7.1}$$

Moreover, let us suppose that $t \longrightarrow w(t)$ is an α-isoptic of a certain curve

$$t \longrightarrow z(t) = p(t)e^{it} + \dot{p}(t)ie^{it}. \tag{7.2}$$

We define two positive-valued functions g, h as follows

$$w(t) = g(t)e^{it} + h(t)ie^{it}. \tag{7.3}$$

By (1.6) we recover the support function p as

$$g(t) = p(t) \tag{7.4a}$$

$$h(t) = \frac{1}{\sin\alpha}p(t+\alpha) - p(t)\cot\alpha. \tag{7.4b}$$

From equations (7.4) we have

$$\int_0^{2\pi} h(t)dt = \frac{1}{\sin\alpha}\int_0^{2\pi} p(t+\alpha)dt - \cot\alpha \int_0^{2\pi} p(t)dt. \tag{7.5}$$

what may be written as

$$\tan\frac{\alpha}{2} = \frac{\int_0^{2\pi} h(t)dt}{\int_0^{2\pi} g(t)dt}. \tag{7.6}$$

Since the curves $t \longrightarrow z(t)$ and $t \longrightarrow w(t)$ have the same Steiner centroids, we get

$$\int_0^{2\pi} g(t)e^{it}dt = \int_0^{2\pi} h(t)e^{it}dt = 0. \tag{7.7}$$

The above considerations lead us to the following result.

Theorem 7.1. If $t \longrightarrow w(t) = g(t)e^{it} + h(t)ie^{it}$ is an α-isoptic of a certain oval $t \longrightarrow z(t)$ of class C^2 with the Steiner centroid at 0, then $g > 0$, $h > 0$, g is the support function of $t \longrightarrow z(t)$,

$$\tan\frac{\alpha}{2} = \frac{\int_0^{2\pi} h(t)dt}{\int_0^{2\pi} g(t)dt}, \quad \int_0^{2\pi} g(t)e^{it}dt = \int_0^{2\pi} h(t)e^{it}dt = 0, \tag{7.8}$$

and

$$h(t)dt = \frac{1}{\sin\alpha} g(t+\alpha) - \cot\alpha g(t). \tag{7.9}$$

Conversely, let $t \longrightarrow w(t) = g(t)e^{it} + h(t)ie^{it}$ be a closed simple curve of class C^2, such that $g > 0$, $h > 0$ and $h(t)dt = \frac{1}{\sin\alpha} g(t+\alpha) - \cot\alpha g(t)$ for α defined by the first formula under (7.8) and let the second formula under (7.8) be satisfief by h and g. Then $t \longrightarrow w(t)$ is the α-isoptic of the oval $t \longrightarrow z(t) = g(t)e^{it} + \dot{g}(t)ie^{it}$.

Let $g(t) = r\cos\frac{\alpha}{2}$, $h(t) = r\sin\frac{\alpha}{2}$ for $t \in [0, 2\pi]$, $\alpha \in (0, \pi)$ fixed. Then $g > 0$, $h > 0$ and the curve $w(t) = r\sin\frac{\alpha}{2}e^{it} + r\cos\frac{\alpha}{2}ie^{it}$ is a circle of radius r. Obviously, we have

$$\frac{1}{\sin\alpha} g(t+\alpha) - \cot\alpha g(t) = r\sin\frac{\alpha}{2} = h(t).$$

Moreover, we have

$$\int_0^{2\pi} g(t)e^{it}dt = \int_0^{2\pi} h(t)e^{it}dt = 0.$$

Thus $t \longrightarrow w(t)$ is the α-isoptic of the circle $t \longrightarrow z(t) = r\cos\frac{\alpha}{2}e^{it}$.

References.

[1] Benko, K., Cieślak, W., Góźdź, S., Mozgawa, W.: On isoptic curves, to appear in Analele Stiint., Iasi.

[2] Bonnesen, T., Fenchel, W.: Theorie der konvexen Körper, Chelsea Publ. Comp., New York, 1948.

[3] Santalo, L.: Integral Geometry and Geometric Probability, Encyclopedia of Mathematics and its Applications, Reading, Massachussets, 1976.
[4] Su, Buchin: Lectures on Differential Geometry, World Scientific Singapore, 1980.
[5] Wunderlich, W.: Contributions to the Geometry of Cam Mechanisms with Oscillating Followers, J. Mechanisms, Vol. 6 (1971), 1-20.
[6] Wunderlich, W.: Kurven mit isoptischem Kreis, Aequationes Math. 5 (1971), 69-79.
[7] Wunderlich, W.: Kurven mit isoptischer Ellipse, Monatshefte Math. 75 (1971), 346-362.

This paper is in final form and no version will appear elsewhere.

Instytut Matematyki UMCS
Pl. M.C. Sklodowskiej 1
20-031 Lublin, Poland.

Generalized Cayley Surfaces

Franki Dillen* and Luc Vrancken**

1. Introduction

In $[NP]_2$ K. Nomizu and U. Pinkall classify the affine surfaces in $\mathbb{R}^3$ satisfying $\nabla^2 h = 0$. More precisely, they prove the following theorem.

Theorem $[NP]_2$ *Let* M^2 *be a nondegenerate surface in* $\mathbb{R}^3$. *Let* ∇ *be the canonical affine connection and let* h *be the second fundamental form (affine metric). If* $\nabla^2 h = 0$ but $\nabla h \neq 0$, *then* M *is congruent to an open subset of the Cayley surface* $z = xy + y^3$ *by an equiaffine transformation of* $\mathbb{R}^3$.

We received a preprint of this paper back in 1988 and tried to extend this result. The first thing we proved is the following theorem.

Theorem 1 *Let* $(M^n,\nabla,\theta) \longrightarrow (\mathbb{R}^{n+1},D,\omega)$ *be a nondegenerate equiaffine immersion. Then* $\nabla^2 h = 0$ *if and only if one of the two following statements holds.*

(i) M *is an open part of a nondegenerate quadric and* ∇ *is the induced connection.*

(ii) M *is flat and* M *is congruent under an equiaffine transformation of* $\mathbb{R}^{n+1}$ *to the graph immersion* $x_{n+1} = F(x_1,\ldots,x_n)$, *where F is a polynomial of degree at most 3 with nonzero Hessian.*

So this result generalizes the Nomizu-Pinkall result in one direction. In this paper we concentrate however on the case $n = 2$ (we prove Theorem 1 only for $n = 2$, but the proof can be adapted to general n). The second author considered the case $n > 2$ in [V]. Although [V] is written some time after this paper, it did already appear, even before $[NP]_2$ did. The present paper was written in 1988. This version is a revised one, written in 1990, after the first author gave a lecture on this topic during the conference on Global Differential Geometry and Global Analysis, held at the Technische Universität Berlin in June 1990. The main part of this paper is dedicated to the proof of the following theorem.

---

* Supported by a Research fellowship of the Research Council of the Katholieke Universiteit Leuven

** Research Assistant of the National Fund for Scientific Research (Belgium).

This article is in final form and no version has appeared or will appear elsewhere.

Theorem 2 *Let* M^2 *be a nondegenerate hypersurface of* $\mathbb{R}^3$, *let* ∇ *be the canonical connection on* M^2 *and let h be the affine metric on* M^2. *Then* $\nabla^k h = 0$ *for* $k \in \{2,3,4,5\}$ *if and only if* M^2 *is flat and equiaffine equivalent to the graph immersion* $z = xy + x^3P(x)$, *where* P *is a polynomial in* x *of degree at most* $k - 2$.

So in particular for $k = 2$, Theorem 2 gives the Nomizu-Pinkall result. For the obvious reasons we would like to call the surfaces, described by $z = xy + x^3P(x)$, where P is any polynomial in x, *generalized Cayley surfaces.*

We would like to point out that slightly related results are obtained in [BNS]. In the mean time, we were able to extend Theorem 2 in $[DV]_2$ to arbitrary k under the additional condition that M^2 is affine minimal, i.e. trace $S = 0$, where S is the affine shape operator. The general problem however remains open for $k \geq 6$, as far as we know.

2. Preliminaries

Let M^n be a manifold with affine connection ∇. Let D denote the standard connection on $\mathbb{R}^{n+1}$. By an affine immersion f: $(M^n,\nabla) \longrightarrow (\mathbb{R}^{n+1},D)$ we mean an immersion for which there exists locally (i.e. in a neighborhood of each point) a transversal vector field ξ such that for all tangent vector fields X and Y to M the following formula holds

$$D_X Y = \nabla_X Y + h(X,Y)\xi. \tag{2.1}$$

We call ξ an affine normal vector field. Clearly h is a symmetric bilinear form. If h is non-degenerate, we say that the immersion is *nondegenerate* and if h is positive or negative definite, we say that the immersion is *convex*. It is clear that these definitions do not depend on the choice of the transversal vector field ξ, nor on the connection ∇ we started with.

Let ω be the parallel volume form on $\mathbb{R}^{n+1}$ given by the determinant, let f: $(M^n,\nabla) \longrightarrow (\mathbb{R}^{n+1},D)$ be an affine immersion and ξ an affine normal vector field, then we can define a volume form θ on M by

$$\theta(X_1,X_2,..,X_n) = \omega(X_1,X_2,..,X_n,\xi). \tag{2.2}$$

However, it is clear that, in general, this volume form doesn't need to be parallel. If this volume form is parallel, we say that the immersion f is an *equiaffine* immersion, and we call ξ an *equiaffine normal vector field.* It is proven in $[NP]_1$ that if M admits a parallel volume form, then the affine normal vector field ξ can be chosen such that this parallel volume form coincides with the induced volume form, which implies that ξ can be chosen equiaffine. Further, it is immediately clear that if ξ_1 and ξ_2 are both equiaffine normal vector fields, then $\xi_1 = c\xi_2$, where c is a constant on M. Therefore the associated bilinear form is, up to a constant factor, uniquely determined. From now on, we will always work with equiaffine normal vector fields. The associated bilinear form is then called the affine second fundamental form. From $[NP]_1$ we also recall the following formulas for an equiaffine normal vector field.

$$D_X\xi = -SX, \tag{2.3}$$

where S is a (1.1) tensor field on M. Notice also that the fact that ξ is equiaffine implies that $D_X\xi$ has no component in the direction of ξ. We call S the affine shape operator. The equations of Gauss, Codazzi and Ricci are then given by

$$R(X,Y)Z = h(Y,Z)SX - h(X,Z)SY, \tag{2.4}$$

$$(\nabla h)(X,Y,Z) = (\nabla h)(Y,X,Z), \tag{2.5}$$

$$(\nabla_X S)(Y) = (\nabla_Y S)(X), \tag{2.6}$$

$$h(X,SY) = h(SX,Y), \tag{2.7}$$

where ∇h is defined by $(\nabla h)(X,Y,Z) = Xh(Y,Z) - h(\nabla_X Y,Z) - h(Y,\nabla_X Z)$.
The higher derivatives of h are defined recursively by

$$(\nabla^{k+1}h)(X_1,X_2,\dots,X_{k+3}) = X_1(\nabla^k h)(X_2,\dots,X_{k+3}) - \sum_{i=2}^{k+3}(\nabla^k h)(X_2,\dots,\nabla_{X_1}X_i,\dots,X_{k+3}).$$

From the Gauss equation (2.4) it follows that if h is nondegenerate and if M^n is flat, then $S = 0$. In this respect we mention the following proposition from $[NP]_1$.

<u>Proposition 1</u> $[NP]_1$ *Suppose that* $f:(M^n,\nabla) \longrightarrow \mathbb{R}^{n+1}$ *is an affine immersion with* $S = 0$. *Then* f *is affine equivalent to the graph immersion of a certain function* $F: M^n \longrightarrow \mathbb{R}$.

A graph immersion of a function F is an immersion $f: M^n \longrightarrow \mathbb{R}^{n+1}$ defined by

$$f(p) = A(p) + F(p)\xi,$$

where A is an affine transformation of M^n into some affine hyperplane α of $\mathbb{R}^{n+1}$ and ξ is a constant vector field that is transversal to α. If we take ξ as affine normal vector field, then f is an affine immersion. We prove the following proposition.

<u>Proposition 2</u> *Let* $f: U \subset \mathbb{R}^n \longrightarrow \mathbb{R}^{n+1}$ *be an affine immersion with* $S = 0$ *and* $\nabla^k h = 0$ *for some* $k \in \mathbb{N}$. *Then* f *is affine equivalent to the graph immersion of a polynomial function* $F: U \longrightarrow \mathbb{R}$ *of degree* $k+1$.

<u>Proof</u> By proposition 1, f is a graph immersion. We can suppose that f(U) is given by $x_{n+1} = F(x_1,x_2,\dots,x_n)$, and that ξ is an affine normal vector field. Then a basis of tan-

gent vector fields is given by $\{\frac{\partial}{\partial x_i} = (0,\ldots 0,1,0,\ldots,0,\frac{\partial F}{\partial x_i})\}_{i=1}^n$. Since $D_{\frac{\partial}{\partial x_j}} \frac{\partial}{\partial x_i} = \frac{\partial^2 F}{\partial x_j \partial x_i}\xi$,

we obtain that $\nabla_{\frac{\partial}{\partial x_j}} \frac{\partial}{\partial x_i} = 0$ and $h(\frac{\partial}{\partial x_i}, \frac{\partial}{\partial x_j}) = \frac{\partial^2 F}{\partial x_j \partial x_i}$. Therefore we obtain that

$$(\nabla^k h)(\frac{\partial}{\partial x_{i_1}}, \frac{\partial}{\partial x_{i_2}}, \ldots, \frac{\partial}{\partial x_{i_{k+2}}}) = \frac{\partial^{k+2} F}{\partial x_{i_1} \partial x_{i_2} \ldots \partial x_{i_{k+2}}}.$$

Hence $\nabla^k h = 0$ if and only if all (k+2)-nd derivatives of F are zero. Thus F is a polynomial of degree at most $k + 1$.

■

Now, we will define the induced connection and the canonical affine normal. This normal was first introduced by Blaschke in [B]. In this paper, we follow the approach of Nomizu. For more details, see [N]. Let M^n be a nondegenerate hypersurface in $\mathbb{R}^{n+1}$. For any choice of transversal vector field ξ, we can define an affine connection ∇ and a bilinear form h by (2.1). M is locally oriented by ξ and induces a metric volume form, given by its determinant which we will denote by ω_h. Then, in [N], it is proved that there exists up to sign a unique choice of ξ such that

(i) $\omega_h = \theta$, where θ is the volume form defined by (2.2),

(ii) the volume form $\omega_h = \theta$ is parallel with respect to ∇.

It is immediately clear that the immersion defined in this way is indeed an equiaffine immersion and ∇ is called the *canonical connection* and ξ is called the *canonical affine normal vector field.* The second fundamental form associated to ξ is called the *affine metric* of M. The canonical connection, as well as the canonical affine normal vector field and the affine metric are equiaffine invariants. If we apply a general affine transformation to $\mathbb{R}^{n+1}$, then we change the volume form of $\mathbb{R}^{n+1}$ and we obtain again a canonical affine normal vector field. We call a ξ canonical if it is canonical w.r.t. some volume form of $\mathbb{R}^{n+1}$. If ξ_1 and ξ_2 are both canonical, then $\xi_1 = c\xi_2$ for some constant $c \in \mathbb{R}$, see for instance $[DV]_1$. Hence they determine the same affine connection ∇. The canonical connection is therefore an affine invariant, as well as the line determined by the canonical affine normal vector field at each point.

It is well known and easy to check that the constant affine normal of a graph immersion associated to a function F is canonical if and only if the hessian determinant of F, H(F) is constant, i.e. $\det\left[\frac{\partial^2 F}{\partial x_j \partial x_i}\right]$ is constant.

If $n = 2$, then we also can deduce the apolarity conditions :

$$\epsilon_1(\nabla h)(e_i,e_1,e_1) + \epsilon_2(\nabla h)(e_i,e_2,e_2) = 0, \tag{2.8}$$

where $\{e_1,e_2\}$ is a basis at a point p of M such that $h(e_i,e_i) = \epsilon_i$ and $h(e_1,e_2) = 0$, where $\epsilon_i \in \{-1,1\}$ and $i \in \{1,2\}$. If M is nonconvex, we can also choose a basis $\{u_1,u_2\}$ such that $h(u_1,u_1) = h(u_2,u_2) = 0$ and $h(u_1,u_2) = 1$. Corresponding to this basis the apolarity conditions become:

$$(\nabla h)(u_1,u_1,u_2) = 0 = (\nabla h)(u_1,u_2,u_2). \tag{2.9}$$

From now on we always assume that $n = 2$.

3. Some auxiliary results and proof of the theorems

Let T be a (0,k)-tensor field on M and let R denote the curvature tensor on M. then, we define a (0,k+2)-tensor field $R\cdot T$ on M by

$$R\cdot T(X_1,X_2,..,X_{k+2}) = -T(R(X_1,X_2)X_3,X_4,..,X_{k+2})$$
$$- T(X_3,R(X_1,X_2)X_4,X_5,..,X_{k+2}) - ... - T(X_3,X_4,..,X_{k+1},R(X_1,X_2)X_{k+2}).$$

Then, we can also define $R^m\cdot T$ inductively by

$$R^m\cdot T = R\cdot(R^{m-1}\cdot T)\ .$$

The following lemma is standard.

Lemma 1 *Let* M *be an equiaffine immersion with second fundamental form* h. *Then*

$$(\nabla^k h)(x_1,x_2,...,x_{k+2}) = (\nabla^k h)(x_2,x_1,x_3,...,x_{k+2}) + (R\cdot(\nabla^{k-2}h))(x_1,x_2,...,x_{k+2}).$$

The identity in lemma 1 is called the *Ricci identity.*

Lemma 2 *Let* M *be an equiaffine surface in* $\mathbb{R}^3$ *with second fundamental form* h. *Then* $\nabla^{2k}h = 0$ *implies that* $R^k\cdot h = 0$ *and that* $R^k\cdot\nabla h = 0$.

Proof First, by applying Lemma 1, we obtain that

$$(R\cdot\nabla^{2k-2}h)(x_1,x_2,x_3,x_4,...,x_{2k+2}) = 0$$

and therefore we also have that

$$(R\cdot\nabla^{2k-2}h)(x_1,x_2,x_4,x_3,...,x_{2k+2}) = 0.$$

Hence, by subtracting these two equations and by applying Lemma 1 once more, we find that

$$(R^2\cdot\nabla^{2k-4}h)(x_1,x_2,x_3,...,x_{2k+2}) = 0.$$

Now let us assume that $(R^m\cdot\nabla^{2k-2m}h) = 0$. Then, we know that the following equations hold for all tangent vectors $x_1,x_2,\ldots,x_{2k+2}$:

$$(R^m \cdot \nabla^{2k-2m}h)(x_1,x_2,\ldots,x_{2m+1},x_{2m+2},\ldots,x_{2k+2}) = 0$$

and

$$(R^m \cdot \nabla^{2k-2m}h)(x_1,x_2,\ldots,x_{2m+2},x_{2m+1},\ldots,x_{2k+2}) = 0.$$

Hence, by subtracting these two equations and by applying Lemma 1 once more, we find that

$$(R^{m+1} \cdot \nabla^{2k-2m-2}h)(x_1,x_2,x_3,\ldots,x_{2k+2}) = 0.$$

By induction, the first part of the lemma is proved. The proof of the second part is similar, starting from the fact that $\nabla^{2k}h = 0$ implies that also $\nabla^{2k+1}h = 0$.

■

The proof of the following lemma is completely similar to the proof of Lemma 2.

<u>Lemma 3</u> *Let* M *be an equiaffine surface in* $\mathbb{R}^3$ *with second fundamental form h. Then* $\nabla^{2k+1}h = 0$ *implies that* $R^{k+1} \cdot h = 0$ *and that* $R^{k+1} \cdot \nabla h = 0$.

The following lemma follows immediately from the skew symmetry of the affine curvature tensor R in its first two components.

<u>Lemma 4</u> *Let* M *be an equiaffine surface in* $\mathbb{R}^3$, *with second fundamental form h. Let* $k \in \mathbb{N}$. *Then is* $R^k \cdot h$ *skew symmetric in the i-th and the (i+1)-st component, where* $i \in \{1,3,5,\ldots,2k-3,2k-1\}$.

Now, we give some technical lemmas, which will enable us to prove the theorems. Let $p \in M$ and let $\{e_1,e_2\}$ be a basis of T_pM such that $h(e_1,e_1) = 1$, $h(e_1,e_2) = 0$, and $h(e_2,e_2) = \epsilon$ where $\epsilon = \pm 1$. Then, we can express the shape operator with respect to this basis in the following way :

$$\begin{cases} Se_1 = \lambda_1 e_1 + \alpha e_2 \\ Se_2 = \beta e_1 + \lambda_2 e_2 \end{cases},$$

and (2.7) then implies that $\beta = \alpha\epsilon$. If we put $R = R(e_1,e_2)$, then we have the following lemma.

<u>Lemma 5</u>

$$(R^{2k} \cdot h)(e_1,e_1) = 2(4\det S)^{k-1}(-\epsilon)^{k-1}(2\alpha^2 + \epsilon\lambda_2(\lambda_2 - \lambda_1)),$$

$$(R^{2k} \cdot h)(e_2,e_2) = 2(4\det S)^{k-1}(-\epsilon)^{k-1}(2\alpha^2\epsilon - \lambda_1(\lambda_2 - \lambda_1)),$$

$$(R^{2k} \cdot h)(e_1,e_2) = -2\alpha(4\det S)^{k-1}(-\epsilon)^{k-1}(\lambda_1 + \lambda_2),$$

$$(R^{2k+1} \cdot h)(e_1,e_1) = 2\alpha(-1)^k\epsilon^{k+1}(4\det S)^k,$$

$$(R^{2k+1}\cdot h)(e_2,e_2) = 2a(-1)^{k+1}\epsilon^k(4\det S)^k,$$
$$(R^{2k+1}\cdot h)(e_1,e_2) = (-1)^k\epsilon^{k+1}(4\det S)^k(\lambda_2-\lambda_1).$$

Proof First, we will prove that $R\cdot h$ has the desired form. Indeed, by using the Gauss equation and the expression for S given above, we find that

$$(R\cdot h)(e_1,e_1) = 2h(Se_2,e_1) = 2\epsilon a,$$
$$(R\cdot h)(e_2,e_2) = -2\epsilon h(Se_1,e_2) = -2a,$$
$$(R\cdot h)(e_1,e_2) = h(Se_2,e_2) - \epsilon h(Se_1,e_1) = \epsilon(\lambda_2-\lambda_1).$$

Hence the last three formulas are satisfied for $k = 0$. Next we will assume that the bottom three formulas are satisfied for some number $k-1$, $k > 0$, and we will show that all 6 formulas of the lemma hold for k. We have that

$$\begin{aligned}(R^{2k}\cdot h)(e_1,e_1) &= (-2)\,(R^{2k-1}\cdot h)(-\epsilon a e_1 - \lambda_2 e_2, e_1)\\ &= 2(\epsilon a\, 2a + \lambda_2\,(\lambda_2-\lambda_1))\,(-1)^{k-1}\,\epsilon^k\,(4\det S)^{k-1}\\ &= 2\,(4\det S)^{k-1}\,(-\epsilon)^{k-1}\,(2a^2 + \epsilon\,\lambda_2\,(\lambda_2-\lambda_1)).\end{aligned}$$

Similarly, we also obtain that

$$(R^{2k}\cdot h)(e_2,e_2) = -2\,(4\det S)^{k-1}\,\epsilon^{k-1}\,(-1)^k\,(2a^2\epsilon - \lambda_1\,(\lambda_2-\lambda_1))$$
$$(R^{2k}\cdot h)(e_1,e_2) = -2\,a\,(-\epsilon)^{k-1}(4\det S)^{k-1}\,(\lambda_1+\lambda_2).$$

Then

$$\begin{aligned}(R^{2k+1}\cdot h)(e_1,e_1) &= (-2)\,(R^{2k}\cdot h)(-\epsilon a e_1 - \lambda_2 e_2, e_1)\\ &= 4\,(\det S)^{k-1}\,a\,(-\epsilon)^{k-1}\,(-2\,\lambda_1\lambda_2 + 2\epsilon a^2)\\ &= 2\,a\epsilon\,(-4\epsilon\det S)^k.\end{aligned}$$

Similarly, we obtain

$$(R^{2k+1}\cdot h)(e_2,e_2) = -2\,a\,(-4\epsilon\det S)^k$$
$$(R^{2k+1}\cdot h)(e_1,e_2) = \epsilon\,(\lambda_2-\lambda_1)\,(-4\epsilon\det S)^k$$

This completes the proof of the lemma. ∎

Lemma 6 *If* $R^m\cdot h = 0$ *at* $p \in M$, *for some* $m \geq 3$, *then there exists a basis* $\{e_1,e_2\}$ *of* T_pM *such that either*

(i) $h(e_1,e_1) = 1$, $h(e_2,e_2) = \epsilon$, $h(e_1,e_2) = 0$ *and* $S = \lambda I$, *or,*

(ii) $h(e_1,e_1) = 1$, $h(e_2,e_2) = \epsilon$, $h(e_1,e_2) = 0$ *and* $\begin{cases} Se_1 = \lambda e_1 \\ Se_2 = 0 \end{cases}$, *or,*

$$(iii)\ h(e_1,e_1) = h(e_2,e_2) = 0,\ h(e_1,e_2) = 1 \text{ and } \begin{cases} Se_1 = \lambda e_2 \\ Se_2 = 0 \end{cases}.$$

Proof It is clear that, in all three cases, $R^m \cdot h = 0$. First, we will assume that $\det S \neq 0$. Then it follows from Lemma 5 that $a = 0$ and $\lambda_1 = \lambda_2$. Thus, we have (i).

Therefore, we may assume that $\det S = 0$. But then we know that there exists an eigenvector u of S with eigenvalue zero. Again there are two possibilities.

a. $h(u,u) = 0$.

In this case, we can find a vector v, such that $h(v,v) = 0$ and $h(u,v) = 1$. Using the equation of Ricci, we then obtain that $h(Sv,u) = h(v,Su) = 0$. Hence Sv has no component in the direction of v. Thus, we have (iii).

b. $h(u,u) \neq 0$.

Here, we may assume, by taking $-\xi$ as normal, that $h(u,u) = 1$. Furthermore, let v be the vector such that $h(v,v) = \epsilon$, $\epsilon = \pm 1$, and $h(u,v) = 0$. By applying again the equation of Ricci, we find that $h(Sv,u) = h(v,Su) = 0$. Hence Sv has no component in the direction of u. Thus, we have (ii). ∎

Lemma 7 *Let* $p \in M$ *and let us assume that the equation (i) of the previous lemma holds at* p. *If* $R^m \cdot \nabla h = 0$ *then* $S_p = 0$ *or* $\nabla h_p = 0$.

Proof. Since (i) of the previous lemma holds, the curvature tensor at the point p is given by

$$R(X,Y)Z = \lambda\,(h(Y,Z)X - h(X,Z)Y),$$

Then, we will first show by induction that the following formulas hold at the point p :

$$\begin{aligned}(3.1)\qquad &(R^{2n+1}\cdot\nabla h)(e_1,e_1,e_1) = (-1)^{n+1}\epsilon^{n+1}\,3^{2n+1}\,\lambda^{2n+1}\,(\nabla h)(e_2,e_2,e_2)\\ &(R^{2n+1}\cdot\nabla h)(e_1,e_2,e_2) = (-1)^{n}\,\epsilon^{n}\,3^{2n+1}\,\lambda^{2n+1}\,(\nabla h)(e_2,e_2,e_2)\\ &(R^{2n+1}\cdot\nabla h)(e_2,e_2,e_2) = (-1)^{n}\,\epsilon^{n}\,3^{2n+1}\,\lambda^{2n+1}\,(\nabla h)(e_1,e_1,e_1)\\ &(R^{2n+1}\cdot\nabla h)(e_1,e_1,e_2) = (-1)^{n+1}\,\epsilon^{n+1}\,3^{2n+1}\,\lambda^{2n+1}\,(\nabla h)(e_1,e_1,e_1),\end{aligned}$$

where $\{e_1,e_2\}$ is a basis of T_pM such that $h(e_1,e_1) = 1$, $h(e_2,e_2) = \epsilon$, $\epsilon \in \{-1,1\}$ and $h(e_1,e_2) = 0$. First, by applying the definition of $R\cdot\nabla h$ and the apolarity condition, we obtain that

$$\begin{aligned}(R\cdot\nabla h)(e_1,e_1,e_1) &= -\,3\,(\nabla h)(R(e_1,e_2)e_1,e_1,e_1)\\ &= 3\,\lambda\,(\nabla h)(e_2,e_1,e_1) = -\,3\,\lambda\,\epsilon\,(\nabla h)(e_2,e_2,e_2).\end{aligned}$$

Similarly, we find that

$$(R\cdot\nabla h)(e_2,e_2,e_2) = 3\,\lambda\,(\nabla h)(e_1,e_1,e_1),$$
$$(R\cdot\nabla h)(e_1,e_1,e_2) = -3\,\lambda\,\epsilon\,(\nabla h)(e_1,e_1,e_1),$$
$$(R\cdot\nabla h)(e_1,e_2,e_2) = 3\,\lambda\,(\nabla h)(e_2,e_2,e_2).$$

Hence (3.1) holds for $n = 0$. Therefore let us now assume that (3.1) holds for a natural number n and prove that (3.1) also holds for $n+1$. Using the skew symmetry of $R^{2n+1}\cdot\nabla h$ in its $(4k+1)$–st and $(4k+2)$–nd components for $k = 1,2,\dots,n$, and the induction hypothesis, we obtain that

$$\begin{aligned}(R^{2n+2}\cdot\nabla h)(e_1,e_1,e_1) &= 3\,\lambda\,(R^{2n+1}\cdot\nabla h)(e_1,e_2,\dots,e_2,e_1,e_1)\\ &= \epsilon^{n+1}\,(-1)^{n+1}\,3^{2n+2}\,\lambda^{2n+2}\,(\nabla h)(e_1,e_1,e_1).\end{aligned}$$

Similarly, we also obtain that

$$(R^{2n+2}\cdot\nabla h)(e_1,e_1,e_2) = \epsilon^{n+2}\,(-1)^{n+2}\,3^{2n+2}\,\lambda^{2n+2}\,(\nabla h)(e_2,e_2,e_2).$$
$$(R^{2n+2}\cdot\nabla h)(e_1,e_2,e_2) = \epsilon^{n+2}\,(-1)^{n+2}\,3^{2n+2}\,\lambda^{2n+2}\,(\nabla h)(e_1,e_1,e_1).$$
$$(R^{2n+2}\cdot\nabla h)(e_2,e_2,e_2) = \epsilon^{n+1}\,(-1)^{n+1}\,3^{2n+2}\,\lambda^{2n+2}\,(\nabla h)(e_2,e_2,e_2).$$
$$(R^{2n+3}\cdot\nabla h)(e_1,e_1,e_1) = \epsilon^{n+2}\,(-1)^{n+2}\,3^{2n+3}\,\lambda^{2n+3}\,(\nabla h)(e_2,e_2,e_2).$$
$$(R^{2n+3}\cdot\nabla h)(e_1,e_1,e_2) = \epsilon^{n+2}\,(-1)^{n+2}\,3^{2n+3}\,\lambda^{2n+3}\,(\nabla h)(e_1,e_1,e_1).$$
$$(R^{2n+3}\cdot\nabla h)(e_1,e_2,e_2) = \epsilon^{n+1}\,(-1)^{n+1}\,3^{2n+3}\,\lambda^{2n+3}\,(\nabla h)(e_2,e_2,e_2).$$
$$(R^{2n+3}\cdot\nabla h)(e_2,e_2,e_2) = \epsilon^{n+1}\,(-1)^{n+1}\,3^{2n+3}\,\lambda^{2n+3}\,(\nabla h)(e_1,e_1,e_1).$$

Hence (3.1) holds for every n. Using the apolarity condition, we find from (3.1) and the assumption of the lemma that the following holds at p :

$$\lambda(p) = 0 \text{ or } (\nabla h)_p = 0.$$ ■

<u>Proof of Theorem 1</u> By Lemma 2, we know that $\nabla^2 h = 0$ implies that $R\cdot h = 0$. Using the formulas for $R\cdot h = 0$, given in Lemma 5, we immediately obtain that $S = \lambda I$, where λ is a differentiable function on M. The Codazzi equation (2.6) then implies that λ is a constant on M.

If ∇h is identically zero, we obtain by the Berwald theorem ($[N\text{-}P]_3$, [D-V]) that M is a part of a nondegenerate quadric and the connection ∇ coincides with the canonical connection. Therefore, we may assume that there is a point p for which $(\nabla h)_p \neq 0$. However, $\nabla^2 h = 0$ also implies by Lemma 2 that $R\cdot\nabla h = 0$. Suppose that $\lambda \neq 0$. Then from Lemma 7 we get a contradiction. Hence $S_p = 0$.

■

Lemma 8 *Let* M *be a surface with induced connection* ∇ *and canonical affine normal* ξ. *Let* $p \in M$ *and assume that condition (ii) of Lemma 6 holds at the point* p. *Then* $R^2 \cdot \nabla h = 0$ *implies that* $S_p = 0$ *or* $(\nabla h)_p = 0$.

Proof We take a basis $\{e_1, e_2\}$ of T_pM which satisfies (ii). Using Lemma 4 and the Gauss equation, we find that

$$0 = (R^2 \cdot (\nabla h))(e_2, e_2, e_2) = -3\epsilon\lambda(R \cdot (\nabla h))(e_1, e_2, e_2) = 6\lambda^2(\nabla h)(e_1, e_1, e_2),$$

and

$$0 = (R^2 \cdot (\nabla h))(e_1, e_2, e_2) = 2\lambda^2(\nabla h)(e_1, e_1, e_1)$$

Let us now assume that S_p is not identically zero. Then, we deduce from these formulas and the apolarity conditions that $(\nabla h)_p = 0$. ∎

The following lemma was proved in the original version of this paper for $k = 5$. The lemma is however true for all k. For the proof, the reader is referred to $[D\text{-}V]_2$.

Lemma 9 *Let* $\nabla^k h = 0$ *and let* $\{E_1, E_2\}$ *be a local orthonormal basis defined on an open set U, such that for each q in U,* $\{E_1(q), E_2(q)\}$ *satisfies condition (iii) of Lemma 6. Then* $S = 0$ *on U.*

Proof of Theorem 2. In all the cases, $\nabla^5 h = 0$. From Lemma 3, it then follows that $R^3 \cdot h = 0$ and $R^2 \cdot \nabla h = 0$. By combining Lemma 6, Lemma 7 and Lemma 8, we then obtain at each point p that either one of the three following statements holds.

(i) $S_p = 0$, or

(ii) $(\nabla h)_p = 0$, or

(iii) The condition (iii) of Lemma 6 holds at the point p.

Let $\Omega = \{p \in M | S_p = 0 \text{ or } (\nabla h)_p = 0\}$. Then Ω is closed in M. Hence $M \backslash \Omega$ is open in M. Let us assume that $M \backslash \Omega$ is not empty. Then, we can take $p \in M \backslash \Omega$. Then, since condition (iii) of Lemma 6 must be fulfilled at the point p, $M \backslash \Omega$ is nonconvex. Hence there exists a basis $\{E_1, E_2\}$ in a neighborhood of p such that $h(E_1, E_1) = h(E_2, E_2) = 0$ and $h(E_1, E_2) = 1$. Since in each point there are only two null directions, we deduce from the fact that condition (iii) holds on $M \backslash \Omega$ that

$$\begin{cases} SE_1 = \lambda E_2 \\ SE_2 = 0 \end{cases},$$

where λ is a differentiable function on a neighborhood of p in $M \backslash \Omega$. Then Lemma 9 implies that $\lambda(p) = 0$. Hence $S_p = 0$. Thus $M \backslash \Omega = \emptyset$.

Let us then define Ω_1 by $\Omega_1 = \{p \in M \mid S_p \neq 0\}$. First, we assume that Ω_1 is not empty. Then Ω_1 is a non empty open part of M and in each point p of Ω_1, we know that $(\nabla h)_p = 0$. The classical Pick-Berwald theorem then implies that Ω_1 is an open part of a nondegenerate ellipsoid or hyperboloid. Thus detS is a constant different from zero on Ω_1. The continuity of detS then implies that $\Omega_1 = M$.

Finally, we may assume that $S = 0$ on the whole of M. Thus by Proposition 2, we can suppose that M is given by the equation $z = P(x,y)$, where P is a polynomial of degree at most $k+1$, and that the canonical affine normal vector field is given by (0,0,1). We now have to express the condition that ξ is canonical. This condition is

$$(*) \qquad \begin{vmatrix} P_{xx} & P_{xy} \\ P_{xy} & P_{yy} \end{vmatrix} = \epsilon, \quad \epsilon = \pm 1.$$

If M is convex, then $\epsilon = 1$, and by a theorem of Jörgens [J], M is a paraboloid, and consequently $\nabla h = 0$. Now suppose M is not convex. The first author showed in $[D]_1$, see also $[D]_2$, that (*) implies that, up to affine transformations in the (x,y)-plane, $P(x,y) = xy + x^3F(x)$, where F is a polynomial of degree at most $k-2$. Hence M is a generalized Cayley surface. ■

Going through the different lemmas, it is easy to see that we in fact proved the following result.

<u>**Theorem 3**</u> *Let* M^2 *be a nondegenerate hypersurface of* $\mathbb{R}^3$, *let* ∇ *be the canonical connection on* M^2 *and let h be the affine metric on* M^2. *If* $\nabla^k h = 0$, *then either*

(1) $k = 1$ *and* M *is a part of a quadric,*

or else

(2) $\det S = 0$.

If moreover $\nabla h \neq 0$ *and one of the conditions*

(a) $\operatorname{tr} S = 0$

or

(b) $2 \leq k \leq 5$

holds, then

$S = 0$ *and* M *is a generalized Cayley surface.*

References

[B] : W. Blaschke, *Vorlesungen über Differentialgeometrie* ,II, Springer, Berlin, 1923.

[BNS]: N. Bokan, K. Nomizu, U. Simon, *Affine hypersurfaces with parallel cubic forms,* Tôhoku Math. J. 42 (1990), 101-108

[C] : E. Calabi, *Hypersurfaces with maximal affinely invariant area*, Amer. J. Math. 104 (1982), 91-126.

[D]$_1$: F. Dillen, *Hypersurfaces of a real space form with parallel higher order fundamental form*, preprint.

[D]$_2$: ———, *Polynomials with constant Hessian determinant*, to appear in : J. Pure Appl. Algebra

[DV]$_1$: F. Dillen and L. Vrancken, *Canonical affine connection on complex hypersurfaces of the complex affine space*, in : Geometry and Topology of Submanifolds, Marseille, France 18-23 May 1987, 1989, World Scientific, Singapore, 98-111.

[DV]$_2$: ———, *Affine minimal higher order parallel affine surfaces*, to appear in : The Problem of Plateau, A tribute to Jesse Douglas and T. Rado.

[J] : K. Jörgens, *Über die Lösungen der Differentialgleichung $rt-s^2 = 1$*, Math. Ann. 127 (1954), 130-134.

[N] : K. Nomizu, *On completeness in affine differential geometry*, Geom. Dedicata 20 (1986), 43-49.

[NP]$_1$: K. Nomizu and U. Pinkall, *On the geometry of affine immersions*, Math. Z. 195 (1987), 165-178.

[NP]$_2$: ———, *Cayley surfaces in affine differential geometry*, Tôhoku Math. J 41 (1989), 589-596.

[V] : L. Vrancken, *Affine higher order parallel hypersurfaces*, Ann. Fac. Sci. Toulouse 9 (1988), 341-353.

Authors' address :

Katholieke Universiteit Leuven,
Faculteit Wetenschappen,
Departement Wiskunde,
Celestijnenlaan 200 B,
B-3001 Leuven,
Belgium.

On a certain class of conformally flat Euclidean hypersurfaces

A. Ferrández[1*] O.J. Garay[2*] P. Lucas[1]**

1 Introduction

Let $x : M^p \longrightarrow \mathbb{R}^{n+1}$ be an isometric immersion of a manifold M^p into the Euclidean space $\mathbb{R}^{n+1}$ and Δ its Laplacian. The family of such immersions satisfying the condition $\Delta x = \lambda x$, $\lambda \in \mathbb{R}$, is characterized by a well-known result of Takahashi, [10]: they are either minimal in $\mathbb{R}^{n+1}$ or minimal in some Euclidean hypersphere. These submanifolds obviously satisfy the condition $\Delta H = \lambda H$, H being the mean curvature vector field in $\mathbb{R}^{n+1}$. Let us write C_λ as the family of those submanifolds satisfying $\Delta H = \lambda H$. Then C_λ contains the Takahashi's family as a proper subfamily as the cylinder $S^p \times \mathbb{R}^q$ shows. When M^p is compact, both families are the same. Therefore, it is interesting to ask for the following geometric question:

> *"Are there any other submanifolds in C_λ apart from cylinders and Takahashi's family?"*

In this context it is worth exploring the existence of non-minimal Euclidean submanifolds whose mean curvature vector be harmonic, i.e., $\Delta H = 0$. As a first stage, only a special case of Euclidean hypersurfaces will be involved in our study. More concretely, we analize the conformally flat hypersurfaces of $\mathbb{R}^{n+1}$ in C_λ. Actually, we show that this class of hypersurfaces is rather small. Essentially they are either minimal, or hyperspheres or right circular cylinders (see Theorem 3.2). Although there have been several attempts of clasifying conformally flat hypersurfaces of $\mathbb{R}^{n+1}$, [7], [8], these results are not complete as Cecil and Ryan show in [3], where they also obtained a classification of the conformally flat hypersurfaces tauthy embedded in $\mathbb{R}^{n+1}$, $n > 3$.

In proving our result, we have used a method developed by B-Y Chen in [6]. As a result, a significant fact is crucial in our computations: A conformally flat hypersurface M^n, $n > 3$, is characterized by having two principal curvatures (not necessarily distinct)

---

*Partially supported by a DGICYT Grant No. PS87-0115
**Supported by a FPPI Grant, Program PG, Ministerio de Educación y Ciencia

of multiplicities 1 and $n-1$ respectively. Thus even in the case of hypersurfaces with two principal curvatures of arbitrary multiplicities the method does not work. However the procedure also holds good in the case of Euclidean surfaces. This fact allow us to classify the Euclidean surfaces in $\mathcal{C}_\lambda$ (Proposition 3.1). This result was also implicitly contained in the proof of main theorem of [6].

2 Basic Lemmas

Let M^n be an orientable hypersurface in the Euclidean space $\mathbb{R}^{n+1}$. Let us denote by σ, A, H, ∇ and D the second fundamental form, the Weingarten endomorphism, the mean curvature vector, the Riemannian connection of M^n and the normal connection of M^n in $\mathbb{R}^{n+1}$. Then following [4, p. 271] we have

Lemma 2.1 *If M^n is a hypersurface in $\mathbb{R}^{n+1}$ then*

$$\Delta H = \Delta^D H + |\sigma|^2 H + Tr(\bar{\nabla} A_H),$$

Δ being the Laplacian of M^n acting on (n+1)-valued functions, $\Delta^D H$ the Laplacian in the normal bundle and $Tr(\bar{\nabla} A_H)$ is the trace of $\bar{\nabla} A_H = \nabla A_H + A_{DH}$.

Now next lemma tell us how to compute $Tr(\bar{\nabla} A_H)$ in a more familiar form.

Lemma 2.2 *Suppose M^n is an orientable hypersurface of $\mathbb{R}^{n+1}$ and ξ a global unit normal vector field. Let α be the mean curvature with respect to ξ, that is $H = \alpha\xi$. Then*

$$Tr(\bar{\nabla} A_H) = \frac{n}{2}\nabla\alpha^2 + 2A(\nabla\alpha),$$

$\nabla\alpha$ being the gradient of α.

Proof. Choose an orthonormal local frame $\{E_i\}_{i=1}^{n+1}$ in such a way that $\{E_i\}_{i=1}^{n}$ are tangent vector fields to M^n and $E_{n+1} = \xi$. Moreover we assume that $\{E_i\}_{i=1}^{n}$ are eigenvectors of $A_\xi = A$ corresponding to the eigenvalues μ_i, $AE_i = \mu_i E_i$, $i = 1, \ldots, n$. Denote by $\{\omega^1, \ldots, \omega^{n+1}\}$ and $\{\omega_i^j\}$, $i, j = 1, \ldots, n+1$, the dual frame and connection forms associated to $\{E_i\}_{i=1}^{n+1}$, respectively. Then, using the connection equations:

$$\nabla_{E_i} E_j = \sum_{k=1}^{n} \omega_j^k(E_i) E_k,$$

we obtain

$$(\nabla_{E_i} A_H) E_j = E_i(\alpha)\mu_j E_j + \alpha E_i(\mu_j) E_j + \alpha \sum_k (\mu_j - \mu_k)\omega_j^k(E_i) E_k.$$

But then by Codazzi's equation

$$0 = \alpha E_i(\mu_j) E_j - \alpha E_j(\mu_i) E_i + \alpha \sum \{(\mu_j - \mu_k)\omega_j^k(E_i) - (\mu_i - \mu_k)\omega_i^k(E_j)\} E_k.$$

Therefore

$$\alpha E_j(\mu_i) = \alpha(\mu_j - \mu_i)\omega_i^j(E_i).$$

Consequently

$$Tr(\bar{\nabla} A_H) = \frac{n}{2}\nabla\alpha^2 + 2\,TrA_{DH},$$

and since $TrA_{DH} = \sum_i A_{D_{E_i}H}E_i = A(\nabla\alpha)$ the lemma follows.

Remark 2.1 We denote by $\mathcal{U} = \{p \in M^n : \nabla\alpha^2(p) \neq 0\}$. $\mathcal{U}$ is an open set in M^n and as a consequence of Lemma 2.2, $Tr\bar{\nabla}A_H = 0$ if and only if $A(\nabla\alpha^2) = -\frac{n}{2}\alpha\nabla\alpha^2$ on $\mathcal{U}$.

Before ending this section we would like to give a first application of Lemma 2.2.

Proposition 2.1 *Let M^n be a compact hypersurface immersed in $\mathbb{R}^{n+1}$. Then M^n has constant mean curvature α if and only if $\Delta H = |\sigma|^2 H$, $|\sigma|^2$ being the length of the second fundamental form.*

Proof. Suppose M^n has constant mean curvature α. Then using Lemma 2.2 we have $Tr(\bar{\nabla}A_H) = 0$. Now take a local orthonormal frame $\{E_i\}_{i=1}^n$ in a neighborhood of a given point p. We can choose the fields $\{E_i\}_{i=1}^n$ in such a way that $\nabla_{E_i}E_j(p) = 0$. Therefore

$$\Delta^D H(p) = -\left(\sum_{i=1}^n D_{E_i}D_{E_i}H - D_{\nabla_{E_i}E_i}H\right)(p) = -\sum_{i=1}^n D_{E_i}D_{E_i}H(p) = \Delta(\alpha)\xi(p).$$

Thus $\Delta^D H = \Delta(\alpha)\xi$ on the whole M^n. Since α is constant, $\Delta^D H = 0$. Consequently we get from Lemma 2.1 $\Delta H = |\sigma|^2 H$.

Conversely, suppose $\Delta H = |\sigma|^2 H$. Then from Lemma 2.1, $\Delta^D H + Tr(\bar{\nabla}A_H) = 0$. Hence if we write normal and tangent components, we have $\Delta^D H = Tr(\bar{\nabla}A_H) = 0$. As before, $\Delta^D H = \Delta(\alpha)\xi$, then $\Delta(\alpha) = 0$. Since M^n is compact, α is constant.

Remark 2.2 Let us denote by τ the scalar curvature of M^n. From the above proposition and Takahashi's theorem, it is easy to prove that the only compact immersed hypersurfaces of $\mathbb{R}^{n+1}$ having constant two of the three following quantities α, $|\sigma|^2$ and τ, are hyperspheres. See also [5].

3 Main Result

Our goal is to prove the following theorem.

Theorem 3.1 *Let M^n be a conformally flat orientable hypersurface of $\mathbb{R}^{n+1}$, $n > 3$. If M^n is in the family C_λ (i.e. $\Delta H = \lambda H$), for a constant λ, then it is either minimal or isoparametric.*

Proof. Suppose M^n is conformally flat in $\mathbb{R}^{n+1}$, $n > 3$. If M^n is totally umbilical, then M^n is a piece of $\mathbb{R}^n$ or S^n. Otherwise, from Theorem 3 of [8] the Weingarten map of M^n has two distinct eigenvalues of multiplicities 1 and $n-1$, respectively. Our next step is to prove that M^n has constant mean curvature. If α were not constant, then by the Remark 2.1 $\mathcal{U}$ is not empty and the vector $\nabla\alpha^2$ is an eigenvector of A corresponding to the eigenvalue $-\frac{n}{2}\alpha$. Choose a local frame $\{E_i\}_{i=1}^{n+1}$, $E_{n+1} = \xi$, in an open set of $\mathcal{U}$ satisfaying that $\{E_i\}_{i=1}^{n}$ are eigenvectors of A and E_1 is parallel to $\nabla\alpha^2$. Thus one of the eigenvalues is $-\frac{n}{2}\alpha$. We have then two possible cases:

a) $-\frac{n}{2}\alpha$ has multiplicity 1, and therefore the other eigenvalue is $\frac{3}{2}\frac{n}{n-1}\alpha$ with multiplicity $n-1$.

b) $-\frac{n}{2}\alpha$ has multiplicity $n-1$ and the other eigenvalue is $\frac{n(n+1)}{2}\alpha$ with multiplicity 1.

Either choice of the multiplicity of $-\frac{n}{2}\alpha$ will lead to the same conclusion, so there is no loss of generality in assuming we are in the first case.

Now by hypothesis $\Delta H = \lambda H$ so that from Lemmas 2.1 and 2.2 we have

$$\Delta^D H = (\lambda - |\sigma|^2)H; \quad A(\nabla\alpha) + \frac{n}{2}\alpha\nabla\alpha = 0. \tag{3.1}$$

Let $\{\omega^1, \ldots, \omega^{n+1}\}$ and $\{\omega_i^j\}_{i,j=1,\ldots,n+1}$ the dual frame and the connection forms of the choosen frame. Then we have

$$\omega_{n+1}^1 = \frac{n}{2}\alpha\omega^1; \qquad \omega_{n+1}^j = -\frac{3}{2}\frac{n}{n-1}\alpha\omega^j, j = 2, \ldots, n. \tag{3.2}$$

$$d\alpha = E_1(\alpha)\omega^1. \tag{3.3}$$

From the first equation of (3.2) we have

$$d\omega_{n+1}^1 = \frac{n}{2}\alpha d\omega^1. \tag{3.4}$$

Using now the second equation of (3.2) and the structure equations, one has

$$d\omega_{n+1}^1 = -\frac{3}{2}\frac{n}{n-1}\alpha d\omega^1. \tag{3.5}$$

These two last equations mean that

$$d\omega^1 = 0. \tag{3.6}$$

Therefore one locally has $\omega^1 = du$, for a certain function u, which along with (3.3) imply that $d\alpha \wedge du = 0$. Thus α depends on u, $\alpha = \alpha(u)$. Then $d\alpha = \alpha' du = \alpha'(u)\omega^1$ and so $E_1(\alpha) = \alpha'$.

Taking differentiation in the second equation of (3.2) we have

$$d\omega_{n+1}^j = -\frac{3}{2}\frac{n}{n-1}\alpha'\omega^1 \wedge \omega^j - \frac{3}{2}\frac{n}{n-1}\alpha d\omega^j, \tag{3.7}$$

and, also by the structure equations:

$$d\omega_{n+1}^j = -\frac{3}{2}\frac{n}{n-1}\alpha d\omega^j - \frac{n(n+2)}{2(n-1)}\alpha\omega_1^j \wedge \omega^1. \tag{3.8}$$

Consequently

$$\omega_j^1 = \frac{3}{n+2}\frac{\alpha'}{\alpha}\omega^j,\ j = 2, \ldots, n, \tag{3.9}$$

that is

$$(n+2)\alpha\omega_j^1 = 3\alpha'\omega^j,\ j = 2, \ldots, n. \tag{3.10}$$

Differentiating (3.10) and using (3.2) and (3.9) we have

$$d(\alpha\omega_j^1) = \frac{3}{n+2}\frac{(\alpha')^2}{\alpha}\omega^1 \wedge \omega^j + \alpha d\omega_j^1, \tag{3.11}$$

$$d\omega_j^1 = -\frac{3}{4}\frac{n^2}{n-1}\alpha^2\omega^1 \wedge \omega^j + \frac{3}{n+2}\frac{\alpha'}{\alpha}(d\omega^j + \frac{3}{n+2}\frac{\alpha'}{\alpha}\omega^1 \wedge \omega^j). \tag{3.12}$$

On the other hand

$$d(\alpha'\omega^j) = \alpha''\omega^1 \wedge \omega^j + \alpha' d\omega^j. \tag{3.13}$$

Hence from (3.10) to (3.13) we obtain

$$4\alpha\alpha'' - \frac{4(n+5)}{n+2}(\alpha')^2 + \frac{n^2(n+2)}{n-1}\alpha^4 = 0. \tag{3.14}$$

Putting $y = (\alpha')^2$ the above equation turns into

$$2\alpha y' - \frac{4(n+5)}{n+2}y = -\frac{n^2(n+2)}{n-1}\alpha^4, \tag{3.15}$$

and then

$$y = (\alpha')^2 = C\alpha^{\frac{2(n+5)}{n+2}} - \left(\frac{n(n+2)}{2(n-1)}\right)^2 \alpha^4, \tag{3.16}$$

with C a constant.

Now we use the definition of $\Delta\alpha$, the fact that E_1 is parallel to $\nabla\alpha^2$ and equation (3.9) to obtain

$$(n+2)\alpha\Delta\alpha = -(n+2)\alpha\alpha'' + 3(n-1)(\alpha')^2. \tag{3.17}$$

As we know $\Delta^D H = (\Delta\alpha)\xi$, hence from (3.1) we get

$$\alpha\Delta\alpha = (\lambda - |\sigma|^2)\alpha^2. \tag{3.18}$$

Since $|\sigma|^2 = \dfrac{n^2(n+8)}{4(n-1)}\alpha^2$, combining (3.17) and (3.18), we have

$$\alpha\alpha'' - \frac{3(n-1)}{(n+2)}(\alpha')^2 + \left(\lambda - \frac{n^2(n+8)}{4(n-1)}\alpha^2\right)\alpha^2 = 0. \tag{3.19}$$

Thus, putting together (3.14) and (3.19) one has

$$\frac{2(4-n)}{n+2}(\alpha')^2 = \frac{n^2(n+5)}{2(n-1)}\alpha^4 - \lambda\alpha^2. \tag{3.20}$$

We deduce, using (3.16) and (3.20) that α is locally constant on $\mathcal{U}$, which is a contradiction with the definition of $\mathcal{U}$. Hence α is constant on M^n. Taking again (3.1) into consideration, we have $(\Delta\alpha)\xi = (\lambda - |\sigma|^2)H$, so that either $\alpha = 0$ and M^n is minimal or $|\sigma|^2 = \lambda$ and therefore $|\sigma|^2$ is constant. But we had at most two different eigenvalues, then because α and $|\sigma|^2$ are constant, such eigenvalues are also constant. We have therefore that M^n is in fact isoparametric.

A classical result of B. Segre [9] states that the isoparametric hypersurfaces in $\mathbb{R}^{n+1}$ are $\mathbb{R}^n$, $S^n(r)$ and $S^p(r) \times \mathbb{R}^{n-p}$, where $S^p(r)$ is the p-sphere of radius r in the Euclidean subspace $\mathbb{R}^{p+1}$ perpendicular to $\mathbb{R}^{n-p}$. On the other hand, if M^n is minimal and conformally flat with $n \geq 4$, a result of Blair, [1], states that $x(M)$ is contained in a catenoid, see also [2]. Taking into account these results and Theorem 3.1 one has the following.

Theorem 3.2 *Let M^n be a complete conformally flat orientable hypersurface of $\mathbb{R}^{n+1}$, $n > 3$. Then M^n is in the family C_λ if and only if it is one of the following hypersurfaces:*
1) a hyperplane $\mathbb{R}^n$,
2) a catenoid,
3) a round sphere $S^n(r)$,
4) a cylinder over a circle $\mathbb{R}^{n-1} \times S^1(r)$,
5) a cylinder over a round $(n-1)$-sphere $\mathbb{R} \times S^{n-1}(r)$.

Prof. Chen kindly pointed out to us that this result can also be considered under the viewpoint of the finite type theory (see [4]). In fact, it can be shown that a Euclidean immersion satisfying $\Delta H = \lambda H$ is either minimal or of infinite type if $\lambda = 0$, and either of 1-type or of null 2-type if $\lambda \neq 0$.

One should observe that conditions $n > 3$ and conformally flat have been used in order to guarantee the existence of at most two distinct eigenvalues of multiplicities 1 and $n-1$. This is automatically satisfied by a surface in $\mathbb{R}^3$. It means that the above computations are also correct in the case of surfaces of $\mathbb{R}^3$ (see [6]). Then one obtains

Proposition 3.1 *Let M^2 be a surface of $\mathbb{R}^3$ in C_λ. Then either M^2 is minimal or it is a piece of one of the following surfaces: a 2-sphere $S^2(r)$ or a right circular cylinder $S^1(r) \times \mathbb{R}$.*

Remark 3.1 From this result we see that the only surfaces of $\mathbb{R}^3$ satisfying $\Delta H = 0$ are the minimal ones.

References

[1] D. BLAIR, *A generalization of the catenoid.* Canadian J. of Math., **27** (1975), 231-236.

[2] M. do CARMO and M. DAJCZER, *Rotation hypersurfaces in spaces of constant curvature.* Transactions of the A.M.S., **277** (1983), 685-709.

[3] T.E. CECIL and P.J. RYAN, *Conformal geometry and the cyclides of Dupin.* Canadian J. of Math., **32** (1980), 767-782.

[4] B.-Y. CHEN, *Total mean curvature and submanifolds of Finite Type.* World Scientific. Singapore and New Jersey, 1984.

[5] B.-Y. CHEN, *Finite type submanifolds and generalizations.* Instituto "Guido Castelnuovo". Roma, 1985.

[6] B.-Y. CHEN, *Null 2-type surfaces in* $\mathbb{R}^3$ *are circular cylinders.* Kodai Math. J., **11** (1988), 295-299.

[7] R.S. KULKARNI, *Conformally flat manifolds.* Proc. Nat. Acad. Sci. U.S.A., **69** (1972), 2675-2676.

[8] S. NISHIKAWA and Y. MAEDA, *Conformally flat hypersurfaces in a conformally flat manifold.* Tohoku Math. J., **26** (1974), 159-168.

[9] B. SEGRE, *Famiglie di ipersuperficie isoparametrische negli spazi euclidei ad un qualunque numero di dimensioni.* Atti Accad. Naz. Lincei Rend. Cl. Sci. Fis. Mat. Natur., **27** (1938), 203-207.

[10] T. TAKAHASHI, *Minimal immersions of Riemannian manifolds.* J. Math. Soc. Japan, **18** (1966), 380-385.

[1] Departamento de Matemáticas. Universidad de Murcia. Campus de Espinardo, 30100 Espinardo, Murcia, Spain.
[2] Departamento de Geometría y Topología. Universidad de Granada. 18071 Granada, Spain.

SELF-DUAL MANIFOLDS WITH NON-NEGATIVE RICCI OPERATOR

Paul Gauduchon

(CNRS-Paris)

Let (M,g) be any Riemannian manifold of dimension $n > 3$. The Riemann curvature tensor R splits into, cf. e.g. [B] Ch.I.

$$R = \widetilde{Ric} + W ,$$

where W is the Weyl tensor, which only depends on the conformal class of the metric g, and the remaining part $\widetilde{Ric}$ is the <u>Ricci operator</u> of g.

Like the curvature R itself, the Ricci operator $\widetilde{Ric}$ is considered as a (symmetric) endomorphism of the vector bundle $\Lambda^2 TM$.

Let h denote the <u>normalized Ricci tensor</u> of g, deduced from the usual Ricci tensor Ric by

$$h = 1/(n-2)[Ric - 1/2(n-1)\, Scal.g] ,$$

where $Scal$ denote the <u>scalar curvature</u> of g (= the trace of Ric).

Then, the Ricci operator $\widetilde{Ric}$ is given by

$$\widetilde{Ric}(X \wedge Y) = h(X) \wedge Y + X \wedge h(Y) , \qquad X,Y \in TM .$$

By definition, the Ricci operator of (M,g) is <u>positive</u> (<u>non-negative</u>) if $\widetilde{Ric}$ is positive (non-negative) as a (symmetric) endomorphism of $\Lambda^2 TM$ (shortly, $\widetilde{Ric} > 0$ or $\widetilde{Ric} \geq 0$).

Equivalently, $\widetilde{Ric}$ is positive (non-negative) iff h is 2-positive (2-non-negative), i.e.

$$h(X,X) + h(Y,Y) > 0 \quad (\text{resp. } \geq 0) ,$$

for any orthonormal pair $\{X,Y\}$.

Let $\lambda_1 \leq \dots \leq \lambda_n$ be the n eigenvalues of the Ricci tensor Ric.

Then, the eigenvalues of $\widetilde{Ric}$ are the $n(n-1)/2$ following

$$1/(n-2)[\lambda_i + \lambda_j - 1/(n-1)\, Scal] , \quad 1 \leq i < j \leq n .$$

Hence, the non-negativity of the Ricci operator $\widetilde{Ric}$ is equivalent to the following <u>pinching</u> condition

$$\lambda_1 + \lambda_2 \geq 1/(n-2) \sum_{j>2} \lambda_j \quad ,$$

(strict inequality for $\widetilde{Ric} > 0$).
In particular, we have the implications

$$\widetilde{Ric} \geq 0 \quad \Rightarrow \quad Ric \geq 0$$

$$\widetilde{Ric} > 0 \quad \Rightarrow \quad Ric > 0 \quad .$$

For $n = 4$, the pinching condition above reduces to

$$\lambda_1 + \lambda_2 \geq 1/2\ (\lambda_3 + \lambda_4) \quad .$$

If, in addition, M is oriented, the Weyl tensor W splits into

$$W = W_+ + W_- \quad ,$$

where W_+ (W_-) is the selfdual (anti-selfdual) part of W .
The Riemannian metric g is <u>selfdual</u> if the Weyl tensor W is selfdual , i.e. if

$$W_- = 0 \quad .$$

Then, we have the following

THEOREM. <u>Let</u> (M,g) <u>be a compact</u> , <u>4-dimensional selfdual Riemannian manifold</u>, <u>with positive scalar curvature</u>.
<u>If the Ricci operator</u> $\widetilde{Ric}$ <u>is non-negative</u> , <u>then</u>
(i) <u>either</u> (M,g) <u>is locally isometric to the product</u> $S^1 \times S^3$ (<u>where</u> S^j <u>denotes the standard sphere of dimension</u> j <u>and unspecified radius</u>),
(ii) <u>or</u> M <u>is simply-connected</u> , <u>homeomorphic to a connected sum</u> $k\,CP^2$ <u>of</u> k <u>copies of the complex projective plane</u> CP^2 , <u>with</u> $0 \leq k \leq 3$.
<u>If the Ricci operator</u> $\widetilde{Ric}$ <u>is positive</u> , <u>then</u> (M,g) <u>is conformally isomorphic to the sphere</u> S^4 <u>or the complex projective plane</u> CP^2 . <u>with their standard conformal structure</u> .

NOTE. In the statement above, we use the convention that, for $k = 0$, $k\,CP^2$ is the sphere S^4 .

PROOF. Let π be natural projection of the so-called <u>twistor space</u> Z of (M,g) onto M.
For each point x of M, the fiber Z_x is the manifold of g-orthogonal complex structures of the tangent space T_xM, inducing the opposite orientation.
Let Θ denote the vertical tangent bundle on the twistor space Z.
For each point J of Z over x (viewed as an automorphism of T_xM), the fiber Θ_J is naturally identified to the following vector space

$$\Theta_J = \{a \in A^-_xM \mid J \circ a = - a \circ J\} \quad ,$$

where A^-M is the vector bundle of anti-selfdual, skew-symmetric endomorphisms of TM over M.
Via this identification, the rank 2 (real) vector bundle Θ admits a natural complex structure $\mathcal{J}$ defined by

$$\mathcal{J}a = J \circ a \quad , \qquad a \in \Theta_J \ .$$

On the other hand, the vector bundle AM of skew-symmetric endomorphisms of TM, as well as its selfdual and anti-selfdual components A^+M and A^-M, admits a natural euclidean structure defined by

$$\langle a,b\rangle = - 1/2 \ \mathrm{trace}(a \circ b) \quad , \qquad a\ b \in A_xM \ , \ x \in M \ ,$$

in such a way that the twistor space Z is naturally identified with the sphere bundle (of radius $\sqrt{2}$) associated to A^-M.
For any J in Z, the restriction of $\langle .,. \rangle$ to Θ_J is clearly compatible with $\mathcal{J}$. We thus obtain on Θ a canonical structure of a Hermitian vector bundle (of rank 1) and, on each fiber Z_x, a canonical Kähler structure (depending on the conformal class $[g]$ only), isomorphic to the complex projective line CP^1 with the Fubini-Study metric of (constant) sectional curvature $1/2$.
We complete $\mathcal{J}$ into an almost-complex structure on Z, also denoted by $\mathcal{J}$, in the following way.
Let H^D be the horizontal distribution on Z induced by the Levi-Civita connexion D of g. Let v^D denote the associated projection from TZ onto Θ.
Then, each vector U in T_JZ can be identified with the pair

$$U = (v^D(U), X) \ ,$$

where $X = \pi_*(U)$ is the projection of U in T_xM, $x = \pi(J)$.

Via this identification, the almost-complex structure $\mathcal{J}$ is defined by

$$\mathcal{J}U = (J \circ v^D(U), JX) \quad .$$

It is easily checked that $\mathcal{J}$ only depends on the conformal class of g, and that $\mathcal{J}$ is integrable iff g is selfdual (cf.[AHS]; more details concerning the twistorial construction in the form above can be found in $[G_1]$).

It follows that, in the considered case, the twistor space Z is a 3-dimensional complex manifold.

For any Hermitian line bundle F on Z, a <u>Chern connection</u> is a Hermitian linear connection on F, whose curvature is $\mathcal{J}$-invariant.

It is well-known that, for any fixed (fibered) Hermitian structure on F, there is a natural bijection between Chern connections and holomorphic structures on F, obtained by identifying a Chern connection with its (0,1)-part.

The Levi-Civita connection D on (M,g) determines a Chern connection ∇ on Θ as follows.

For any section ξ of Θ, regarded as a (vertical) vector field on Z, we put, for each x on M, each J on Z_x, each U in $T_J Z$,

$$(1) \qquad \nabla_U \xi = \nabla^{(x)}_{v^D(U)} \xi + [\tilde{X}, \xi]_J$$

where

$\nabla^{(x)}$ denotes the Levi-Civita connection of the canonical Kähler structure of the fiber Z_x,

$\tilde{X} = U - v^D(U)$ is the horizontal lift of the projection X of U in $T_x M$, extended, in an arbitrary way, into a horizontal vector field in the neighbourhood of J,

$[\tilde{X}, \xi]_J$ is the value at J of the braket of the vector fields $\tilde{X}$ and ξ.

It is easily checked that $[\tilde{X}, \xi]_J$ only depends upon the <u>vector</u> U, and that (1) defines a linear connection on Θ.

Since the Levi-Civita connection D preserves the metric g, the parallel transport on the fiber bundle Z associated to D preserves the Kähler structure of the fibers, so that ∇ is a Hermitian connection.

The bundle Θ has been realized as a subbundle of $\pi^*(A^-M)$, reciprocal image of A^-M by π. Denoting by π^*D the induced connection on $\pi^*(A^-M)$ (where D is viewed as a connection on A^-M), and by Π the orthogonal projection of $\pi^*(A^-M)$ onto Θ, we check that ∇ is also equal to

$$(2) \qquad \nabla = \Pi \circ \pi^*D \ .$$

Then, it can be shown, by using the expression (1) or (2) of ∇ (cf. resp. $[G_2]$ and $[G_1]$), that the curvature R^∇ of ∇ is J-invariant (= ∇ is a Chern connection) iff g is selfdual.
In this case, the corresponding Chern form $\gamma^D = 1/2\pi i . R^\nabla$ admits the following expression (cf. $[G_1]$ LEMME 5) at the point J of Z_x :

$$(3) \qquad \begin{aligned} \gamma^D_{|\Theta_J} &= 1/4\pi \, \Omega^{(x)}_J \\ \gamma^D_J(\tilde{X}, a) &= 0 \\ \gamma^D_J(\tilde{X}, \tilde{Y}) &= 1/2\pi[h(JX,Y) - h(X,JY)] \ , \end{aligned}$$

where $\Omega^{(x)}$ denotes the Kähler form of the fiber Z_x, $\tilde{X}$ and $\tilde{Y}$ are the horizontal lift, at J, of any X and Y in T_xM, a is any element in Θ_J.

From (3), we infer immediately the following implications

$$(4\text{ a}) \qquad \tilde{Ric} > 0 \qquad \gamma^D > 0 \ ,$$

$$(4\text{ b}) \qquad \tilde{Ric} \geq 0 \qquad \gamma^D \geq 0 \ .$$

In addition, in the latter case, it is easily checked that γ^D is positive somewhere, except in the case when λ_1 vanishes everywhere, i.e., at each point x of M, the eigenvalues of the Ricci tensor are of the form

$$(5) \qquad \lambda_1 = 0 \ , \qquad \lambda_2 = \lambda_3 = \lambda_4 \geq 0 \ .$$

In the case (4 a), it follows, by the well-known KODAIRA criterion, that the twistor space Z is a projective manifold.
Then, by a celebrated theorem of N.J.HITCHIN [H], cf.also [F-K], the <u>conformal</u> manifold $(M,[g])$ has to be isomorphic to the standard S^4 or CP^2.

In case (4 b), we have to distinguish the case when (5) holds everywhere, and the case when λ_1 is positive at some point of M.
In the latter case, we can use the SIU criterion [S] to infer that the twistor space Z is then a Moishezon manifold (= bimeromorphically equivalent to a projective manifold).
We then conclude, by a theorem due to F.CAMPANA [C] that M is simply-connected.
On the other hand, since g is selfdual and its scalar curvature is positive, the intersection form of M is positive, cf. [B] ch.13.
It then follows from the deep works of M.FRIEDMAN and S.DONALDSON that M is actually homeomorphic to $k\,CP^2$, for some $k \geq 0$, cf. [F-U].
Finally, we have, cf. [H],

$$\int_Z (\gamma^D)^3 = 2(2\chi - 3\tau) \quad ,$$

where χ and τ denote respectively the Euler-Poincaré characteristic and the signature of M (equal, respectively, in the present case, to $2 + b_2$ and b_2, where b_2 is the second Betti number of M).
Since γ^D is non-negative, positive somewhere, we infer that b_2 is lesser than 4, then $0 \leq k \leq 3$.

NOTE. SIU criterion asserts that the existence of a Hermitian holomorphic line bundle F on a compact complex manifold Z, whose Chern form is non-negative, positive at some point, implies Z being Moishezon. This result is also an immediate consequence of a stronger criterion deduced from the celebrated "holomorphic Morse inequalities" of J.P.DEMAILLY, cf. [D].

In the case where (5) holds everywhere, the inequality has to be strict at any point, because the scalar curvature is assumed to be positive everywhere. Then, the kernel of the Ricci tensor determines a well-defined real line sub-bundle K of TM.
It follows that the Euler-Poincaré characteristic χ is equal to zero, hence the first Betti number b_1 is positive. By HODGE theory, there exists a non-trivial harmonic 1-form on M, which, by the well-known BOCHNER theorem, has to be parallel, actually a section of K when considered as a (parallel) vector field.
Then, the metric g splits locally into the product $S^1 \times M_o$, where M_o is a 3-dimensional Einstein manifold with positive scalar curvature (equal to 3λ), hence locally isometric to the standard sphere S^3 (of radius $\sqrt{2/\lambda}$). $\square$

REMARK. Selfdual Riemannian conformal structures , for which the twistor space Z is Moishezon , have been constructed in a explicit way on $k\,CP^2$, cf. [P] for k=2 , [LB] for any k . All these conformal structures contain metrics with positive svalar curvature , but it is still unknown (to the author) whether they contain metrics with non-negative Ricci operator .

REFERENCES.

[A-H-S] M.F.ATIYAH-N.J.HITCHIN-I.M.SINGER. Self-duality in four-dimensional Riemannian geometry.Proc.R.Soc.Lon.A.1978,452-461.

[B] A.L.BESSE. Einstein manifolds. Springer Verlag. 1987.

[C] F.CAMPANA. On Twistor Spaces of the Class C . Preprint.1989.

[D] J.P.DEMAILLY. Champs magnétiques et inégalités de Morse pour la d"-cohomologie. C.R.Acad.Sc.Paris,t.301,Série I ,n°4,1985, 119-122.

[F-K] T.FRIEDRICH-H.KURKE. Compact four-dimensional self-dual Einstein manifolds with positive scalar curvature. Math. Nachr.106,1982,271-299.

[F-U] D.S.FREED-K.K.UHLENBECK. Instantons and Four-Manifolds. Springer. MSRI.1984.

[G_1] P.GAUDUCHON. Structures de Weyl et théorèmes d'annulation sur une variété conforme autoduale. Preprint.1989.

[G_2] P.GAUDUCHON. Weyl structures on a selfdual conformal manifold. To appear in Proceedings of the AMS Summer Institute at Los Angeles.1990.

[H] N.J.HITCHIN. Kählerian twistor spaces.Proc.R.Soc.Lond.43, 1981,133-150.

[LB] C.LEBRUN. Explicit Selfdual Metrics on $CP^2 \# \ldots \# CP^2$.Preprint.1990.

[P] Y.S.POON. Compact selfdual manifolds with positive scalar curvature. J.Diff.Geom.24,1986,97-132.

[S] Y.T.SIU. A vanishing theorem for semi-positive line bundles over non-Kähler manifolds. J.Diff.Geom.19,1984,431-452.

On the obstruction group to existence of Riemannian metrics of positive scalar curvature

Bogusław Hajduk

1. Introduction.

The existence of Riemannian metrics of positve scalar curvature on a closed spin manifold M is closely related to vanishing of arithemtic genus and its generalizations. The discovery of this fact goes back to Lichnerowicz [L]. He proved that on such a manifold there is no harmonic spinor, thus in particular its arithmetic genus vanishes. Later Hitchin [Hi] extended this result to the generalized arithmetic genus with values in $KO_*(point)$.

The surgery theorem of Gromow, Lawson [GL1] and Schoen, Yau [SY] revealed the topological aspect of the problem. It says that a metric of positve scalar curvature can be extended through a handle of codimension greater than two (cf. section 2). As a direct consequence one gets that for a spin manifold M, the existence problem depends only on the spin cobordism class of M in $\Omega_*^{Spin}(B\pi_1 M)$ (where BG denotes the classifying space of the group G). A general conjecture formulated by Gromov and Lawson [GL2] states that a closed spin manifold M has a metric of positive scalar curvature if and only if the appropriately defined arithmetic genus of M vanishes on the cobordism class of M in $\Omega_*^{Spin}(B\pi_1 M)$. In such generality the conjecture is known only for simply connected spin manifolds, where the arithmetic genus is the classical one. The fact that vanishing of $\hat{A}$-genus implies existence of a metric of positive scalar curvature for simply connected spin manifolds of dimension greater than four was proved recently by Stephan Stolz [ST] (cf. also [K] and [GL1] for the rational case). There are some results for manifolds with finite fundamental groups in [KS], [R2], [R3].

The particular case of the conjecture we will try to understand is the rational one.

1.1. Rational Gromov-Lawson conjecture. Let M be a closed spin manifold, $\pi = \pi_1 M$. The connected sum of some finite number of copies of M admits a metric of positive scalar curvature if and only if the rational arithmetic genus $\hat{a}_{\mathbf{Q}} : \Omega_*^{Spin}(B\pi) \otimes \mathbf{Q} \longrightarrow KO_*(B\pi) \otimes \mathbf{Q}$ vanishes on the cobordism class of M.

See [Mi2] for the definition of the generalized arithmetic genus $\hat{a}$. For simply connected spin manifolds, the "only if" part of this conjecture is just the Lichnerowicz-Hitchin theorem. The part "if" was proved by Gromow and Lawson [GL1], and their argument extends to arbitrary fundamental group (this extension is due to Matthias Kreck). The conjecture is known to be true for hyperbolic groups in sense of Gromov (see [CM]).

In [H1] we deduced from the surgery theorem that one can define the obstruction to existence of metrics with positive scalar curvature on a spin manifold M in geometric terms. Given a handle decomposition of M, there exists such a metric on $N =$ complement of handles of indices greater than $m-3$, $m = dimM$. By restriction it gives an element $\delta[M]$ in the set of concordance classes of metrics with positive scalar curvature on ∂N. In the simply connected case we can assume $\partial N = S^{m-1}$, thus the obstruction group consists of concordance classes of metrics on S^{m-1}. Moreover, the concordance class $\delta[M]$ is zero if and only if the manifold M admits a metric of positive scalar curvature (cf.[H1]).

In the present paper we give the precise definition of the geometric obstruction group $P_n(\pi, \alpha)$ for arbitrary fundamental group π depending on its finite presentation α. Our construction is functorial with respect to group homomorphisms. Presentations play a role similar to base points in homotopy theory, for instance for any α, α' there is a canonical isomorphism $P_n(\pi, \alpha) \longrightarrow P_n(\pi, \alpha')$. Then we study the obstruction map $\delta_\pi : \Omega_*^{Spin}(B\pi) \longrightarrow P_*(\pi)$. In particular we prove its additivity: $\delta_{\pi * \pi'} = \delta_\pi \times \delta_{\pi'}$, and conclude from this that the class of groups satisfying the rational Gromov-Lawson conjecture is closed with respect to the free product.

A part of this work was done in Bures-sur-Yvette during a stay at IHES. I would like to thank for the invitation and hospitality I found there.

2. Surgery on positive scalar curvature manifolds and a uniqueness lemma.

For a Riemannian metric with positve scalar curvature on a manifold M with boundary we will always assume that on a collar of the boundary, $\partial M \times [0,1]$, it is the product of a metric on ∂M by the standard metric on the interval $I = [0,1]$. In particular, any such metric induces on each component of ∂M a metric of positive scalar curvature. Denote by $PSC(M)$ the set of all metrics of positive scalar curvature on M.

A *concordance* of two metrics $g_0, g_1 \in PSC(M)$, where M is a closed manifold, is a metric of positive scalar curvature on $M \times I$ which induces g_i on $M \times \{i\}$, $i = 0, 1$.

The following theorem gives a construction of metrics of positve scalar curvature related to the topological structure of the manifold and thus allows to interpret a large part of the existence problem in topological terms. It was proved in [GL1], [SY] and we state here an improved version of [G].

2.1. THEOREM. Given a metric g of positve scalar curvature on a closed manifold M and a surgery of codimension greater then 2 on M, there exists an extension of g to a metric of positive scalar curvature (product near the boundary) on the trace of the surgery.

For an account of surgery and handle decompositions see [Mi2]. We collect below some basic definitions we use in the sequel.

A *surgery of codimension* k on M along an embedding $f : S^{m-k} \times D^k \longrightarrow int(M^m)$ is the operation associating to M a new manifold

$$M' = (M^m \setminus f(S^{m-k} \times D^k)) \cup_f D^{m-k+1} \times S^{k-1}.$$

In particular, the connected sum $M \sharp M'$ results from a surgery of codimension m on the disjoint sum $M \cup M'$. We will denote by $g \sharp g'$ the metric on the connected sum resulting from the surgery on the Riemannian manifold $(M \cup M', g \cup g')$.

The manifold

$$M \times I \cup_f D^{m-k+1} \times D^k$$

with boundary $M \cup M' \cup \partial M \times I$ is called the *trace* of the surgery. The disc $D^{m-k+1} \times D^k$ is a *handle of index m-k+1* on $M \times I$ (or (m-k+1)-handle) and f is the *attaching map* of the handle. Let N be a connected component of ∂M. By a *handle decomposition* on N we mean a sequence of submanifolds of codimension O

$$N \times I = M_{-1} \subseteq M_0 \subseteq \ldots \subseteq M_m = M$$

such that M_i is obtained from M_{i-1} by attaching handles of index i, where $0 \leq i \leq m$. Two handles h_1, h_2 are *complementary* if the trace of the corresponding two surgeries is product, i.e. $M \cup h_1 \cup h_2 = M \times I$. Thus as the result of two complementary surgeries we get the manifold we started with.

Consider the disjoint sum of two oriented manifolds W, W' with connected boundaries M, M', and a 1-handle on the sum attached by an embedding which sends the components of $S^0 \times D^{n-1}$ to different components of $M \cup M'$ and preserves orientations. The trace of the corresponding surgery is called *boundary connected sum* of W and W', denoted $W \amalg W'$. It yields a cobordism between $M \cup M'$ and $M \sharp M'$.

Moreover, theorem 2.1. gives a canonical way of extending any given metric to a handle (cf. [GL1]). One can see that if we pass a pair of complementary handles, then the new metric obtained from the two surgeries can be deformed through metrics of positve scalar curvature to the old one. A straightforward corollary of this theorem is the following (cf.[H1], section 2).

2.2. LEMMA. Let X_0, X_1 be connected closed manifolds, W a cobordism between X_0 and X_1 with a handle decomposition on X_0 having no handles of codimension ≤ 2, $g, g' \in PSC(W)$. If $g|X_0 = g'|X_0$, then $g|X_1$ concordant to $g'|X_1$. The codimension assumption can be achieved by a change of the cobordism rel $X_0 \cup X_1$ if W is a spin manifold of dimension ≥ 5 and the inclusion $X_1 \subseteq W$ induces an isomorphism of fundamental groups.

Recall that a relative cobordism rel A between two pairs (W_0, A) and (W_1, A) is a manifold with boundary $W_0 \cup W_1 \cup (A \times I)$, where $A \times \{i\}$ is identified with the copy of A in W_i, $i = 0, 1$. Lemma 2.2. gives in turn the following uniqueness up to concordance of metrics which extend to a spin manifold.

2.3. PROPOSITION. Let X be a connected closed spin manifold. Suppose that W is a compact connected spin manifold of dimension $n \geq 5$ with boundary $X = \partial W$, the inclusion induces an isomorphism $\pi_1 X \longrightarrow \pi_1 W$, and $g \in PSC(X)$ extends to $\tilde{g} \in PSC(W)$. If (W, X, g), (W', X, g') are two such triples and there exists a relative spin cobordism V rel X between (W, X) and (W', X), then g is concordant to g'. In the simplest case, if $g, g' \in PSC(X)$ and both extend to metrics of positive scalar curvature on W, then g is concordant to g'.

Proof. Under the assumption on π_1 there exists a cobordism V' rel X which has no handles of codimension ≤ 2. Following the argument of [H1], Proposition 2.1, the cobordism yields another cobordism (relative to the boundary) of $W \sharp (-W')$ to $X \times I$ without handles of codimension ≤ 2. Thus 2.1 produces the required concordance from the metric $\tilde{g} \sharp \tilde{g}'$ on $W \sharp W'$.

Another easy consequence of 2.2 is that the concordance class of a metric given by Theorem 2.1 is preserved if one changes the attaching map of the handle by an isotopy.

2.4. LEMMA. Let W_1, W_2 be traces of surgeries along isotopic embeddings $f_i : S^k \times D^{n-k} \longrightarrow M$, for $i = 1, 2$, $n - k \geq 3$, and M_1, M_2 be manifolds obtained from M by these surgeries. Suppose that $g \in PSC(M)$ extends to $g_i \in PSC(W)$, $i = 1, 2$. Then there exists a diffeomorphism $F : W_1 \longrightarrow W_2$ such that $F|M = id$ and $(F|M_1)^*(g_2|M_2)$ is concordant to $g_1|M_1$.

3. Thickenings of presentations.

Denote by $(a_1, \ldots, a_k)$ the free group freely generated by $\{a_1, \ldots, a_k\}$; we will identify it with the set of words in $a_1, \ldots, a_k$. A presentation of a group π with generators $a_1, \ldots, a_k$ and relators $r_1, \ldots, r_u \in (a_1, \ldots, a_k)$ will be denoted $(a_1, \ldots, a_k; r_1, \ldots, r_u)$. Thus we have an epimorphism $(a_1, \ldots, a_k) \longrightarrow \pi$ and its kernel is the normal subgroup generated by $r_1, \ldots, r_u$. We allow some relators to be the unit (empty) word. We will consider only finite presentations.

Let α be a presentation. There exists a finite 2-dimensional complex whose 1-cells are in 1-1 correspondence with generators, 2-cells correspond bijectively to relators of α and an attaching map of each 2-cell is given by the corresponding relator. Namely, if $V = S^1 \vee \ldots \vee S^1$ is the wedge of k copies of S^1, then $\pi_1 V$ is the free group on k generators which we identify with $a_1, \ldots, a_k$. For any word r there is an attaching map $r : (S^2, *) \longrightarrow (V, *)$ representing the class in $\pi_1 V$ corresponding to r. The attaching map can be given canonically as a piecewise linear map on the division of S^1 into $t =$ length of r equal intervals. Form a 2-complex $X(\alpha)$ by gluing of 2-cells $D_1, \ldots, D_u$ to V using attaching maps $r_1, \ldots, r_u$. For any $n \geq 5$, embed $X(\alpha)$ to $\mathbf{R}^n$ and let $D_n(\alpha)$ denote a compact regular neighborhood of $X(\alpha)$. Thus $D_n(\alpha)$ is a smooth compact n-manifold with a prescribed trivialization of the tangent bundle. In particular, it is a spin manifold. The fundamental group $\pi_1 D_n(\alpha)$ is canonically identified with π by the identification ψ_α which associates to a_i its image in π. Thus there is a map $\phi_\alpha : D_n(\alpha) \longrightarrow B\pi$, where $B\pi$ denotes the classifying space of π, such that $\phi_{\alpha *} = \psi_\alpha$ on the fundamental group. This map is unique up to homotopy. It is also easy to check that the inclusion $\partial D_n(\alpha) \longrightarrow D_n(\alpha)$ induces an isomorphism of fundamental groups. $D_n(\alpha)$ is canonically decomposed to handles of indices 0,1,2 with the core discs forming the cells of $X(\alpha)$. We say that $D_n(\alpha)$ is a *thickening* of α. Note that $D_n(\alpha)$ may change only in a restricted way when α runs over presentations of π. As one readily sees from the Tietze theorem (cf.[Ha]), two thickenings of π can differ up to a diffeomorphism only by a boundary connected sum with a finite number of copies of $S^2 \times D^{n-2}$. This corresponds to addition (or removal) of superfluous relators (relations which are consequences of remaining relations). Therefore, if α, α' are two presentations of π, then $D_n(\alpha)$ is cobordant to $D_n(\alpha')$ by a cobordism which has all handles of indices 3 or n-3.

4. An interpretation of $\Omega^{Spin}(B\pi)$.

The idea of the construction given in the next section is inherited from the bordism group of the classifying space $B\pi$. To clarify this connection we will present now the related description of $\Omega_n^{Spin}(B\pi)$. This group can be interpreted as spin bordism classes of pairs $\{M, f\}$, where M is a connected spin manifold, and $f : M \longrightarrow B\pi$ is a map such that $f_* : \pi_1 M \longrightarrow \pi_1 B\pi = \pi$ is an isomorphism, with the same assumptions on bordisms. The following well known lemma (cf. [R1]) or rather the proof we present, shows a way to find in any cobordism a representative of this kind. It also yields the adjustments of the disjoint sum juxtaposition to one which associates to any two classes a representative satisfying our assumptions. Compare this with the definition of P_f of section 4.

3.1. LEMMA. Let $n > 4$. Any class of $\Omega_n^{Spin}(B\pi)$ admits a representative (M, f) such that M is a connected spin manifold, $f_* : \pi_1 M \longrightarrow \pi$ is an isomorphism, and $\pi_2 M = 0$. If (M_1, f_1) satisfies this condition and (M_2, f_2) is another representative of the same bordism class, then there exists a spin cobordism (W, F) between (M_1, f_1) and (M_2, f_2) such that $F_* : \pi_1 W \longrightarrow \pi$ is an isomorphism and $\pi_i(W, M_1) = 0$ for $i = 1, 2$.

Proof. For any presentation $\alpha = (b_1, \ldots, b_l; r_1, \ldots, r_u)$ of π there is a map $\phi : D_n(\alpha) \longrightarrow B\pi$ which induces an isomorphism on π_1. Then ϕ extends to f_0 on the double $D_n(\alpha) \cup D_n(\alpha) = M_0$ by composing with the canonical folding map $M_0 \longrightarrow D_n(\alpha)$. Obviously, the resulting class $[M_0, f_0]$ is zero. For a non-trivial class $x = [M, f]$ consider first the connected sum $(M \sharp M_0, f \sharp f_0)$ and kill $\pi_1 M$ by surgery in the following way. Let $\{a_1, \ldots, a_k\}$ be a set of generators of $\pi_1 M$, w_i be a word in $b_1, \ldots, b_l$ representing $(f_0^{-1})_* f_*(a_i)$. Then represent w_i by an embedded loop in $D_n(\alpha)$ and the same for a_i in M. Add now to $(M \sharp M_0) \times I$ 2-handles with attaching maps in the homotopy classes of $a_i w_i^{-1}$ and such that the trace of these surgeries is a spin manifold. In this way we get a spin cobordism between $M \sharp M_0$ and a manifold M'. We have $(f \sharp f_0)_*(a_i w_i^{-1}) = 0$, hence $f \sharp f_0$ extends to the cobordism, and we come to $(M', f') \in x$ with f' inducing an isomorphism in π_1. Since M' is spin, by surgeries of index 3 we can make π_2 trivial. For the remaining part of the lemma one needs only to repeat the argument in the relative case (W, M_1), where (W, F) is arbitrary bordism between the two given representatives.

Note the following special case. Let $h : \pi \longrightarrow \pi'$ be a homomorphism, $[M, f] \in \Omega_n^{Spin}(B\pi)$, $f_* : \pi_1 M \longrightarrow \pi$ an isomorphism. Then the required representative of $[M, Bh \circ f]$ is obtained from $M \sharp M_0$ by surgery as explained above.

The same procedure gives in any bordism class a manifold M' with a handle decomposition which determines any prescribed presentation of π.

5. The functor P.

Let α be a finite presentation of π, $D_n(\alpha)$ its thickening, $n > 5$. Denote by $P_n(\alpha)$ the set of concordance classes of metrics with positve scalar curvature on $\partial D_n(\alpha)$. We will write sometimes $P_n(\pi, \alpha)$ to exhibit the group π. In this section we define a juxtaposition in $P_n(\alpha)$ and we prove functoriality of P_n. Let us describe first a basic geometrical construction.

Let $\alpha' = (a'_1, \ldots, a'_l; r'_1, \ldots, r'_s)$ be a (finite) presentation of another group π', $f : \pi \longrightarrow \pi'$. We assume that f is either a homomorphism or antihomomorphism (i.e. $f(ab) = f(b)f(a)$). For every generator a_i of α, $f[a_i] = [w_i]$, where w_i is a word in $a'_1, \ldots, a'_l$. These choices determine (uniquely up to an isotopy) embeddings

$$a_i w_i^{-1} : S^1 \longrightarrow \partial D_n(\alpha) \sharp \partial D_n(\alpha')$$

with images disjoint from 2-handles of $D_n(\alpha)$ and $D_n(\alpha')$. The trivializations of the tangent bundles of $D_n(\alpha)$ and $D_n(\alpha')$ yield trivializations of normal bundles of the embeddings. By attaching 2-handles to the boundary connected sum $D_n(\alpha) \amalg D_n(\alpha')$ along the resulting embeddings of $S^1 \times D^{n-2}$ we get a spin manifold X_1. Note that a change of the word w_i representing $f[a_i]$ changes the attaching map to an isotopic one, thus does not alter X.

The choice of w_i's induces a homomorphism

$$\phi : (a_1, \dots, a_k) \longrightarrow a'_1, \dots, a'_l).$$

Every relator of α is transformed by ϕ to an element of the normal subgroup generated by relators of α'. This means that in X_1 every attaching map of a 2-handle of $D_n(\alpha)$ is isotopic to a trivial one (the standard embedding in a disc). After such isotopy, any of those handles gives a boundary connected sum with $S^2 \times D^{n-2}$. As every trivially attached handle it can be removed by surgery, i.e. addition of a 3-handle to X_1 attached along $S^2 \times D^{n-1} \subseteq \partial(S^2 \times D^{n-2})$. Let X_2 denote the resulting manifold. Each surgery step was a cancellation of a handle of $D_n(\alpha)$, i.e. we formed a complementary pair of handles with their sum diffeomorphic to a disc (we omit discussing some edges we need to smooth). Thus we get the following.

5.1. LEMMA. X_2 is diffeomorphic to $D_n(\alpha')$ and the trace of the sequence of surgeries is a cobordism W from $\partial D_n(\alpha) \sharp \partial D_n(\alpha')$ to $\partial D_n(\alpha')$ with handles of indices 2 and 3. Both X_2 and W are up to diffeomorphism determined by the map f.

Now we associate to any pair of metrics in $P_n(\alpha) \times P_n(\alpha')$ a new metric in $P_n(\alpha')$ obtaining a map $P_f : P_n(\alpha) \times P_n(\alpha') \longrightarrow P_n(\alpha')$ with nice properties. Let $g \in PSC(\partial D_n(\alpha))$, $g' \in PSC(\partial D_n(\alpha'))$. By 2.1 there is a metric of positve scalar curvature on W extending $g \sharp g'$ and the resulting metric g_0 on $\partial X_2 = \partial D_n(\alpha')$ is unique up to concordance by 2.3 and 2.4. Put $P_f([g],[g']) = [g_0] \in P_n(\alpha')$. The main property of the operation P is associativity in the following sense.

5.2. PROPOSITION. Let $h : \pi \longrightarrow \pi'$, $k : \pi' \longrightarrow \pi''$ be homomorphisms or antihomomorphisms, $\alpha, \alpha', \alpha''$ be presentations of π, π', π'' respectively, $x \in P_n(\alpha)$, $x' \in P_n(\alpha')$, $x'' \in P_n(\alpha'')$. Then

$$P_k(P_h(x, x'), x'') = P_{kh}(x, P_k(x', x'')).$$

Proof. If we start from representatives g, g', gg'' of x, x', x'', then metrics representing both sides of the equality extend $g \sharp g' \sharp g'' \in PSC(\partial D_n(\alpha) \sharp \partial D_n(\alpha') \sharp \partial D_n(\alpha''))$ It is enough to check that traces of surgeries performed in the two cases are cobordant mod boundary. Actually they are diffeomorphic because the surgeries differ only by alterations of attaching maps by isotopies. The attaching maps of 2-handles used to get $P_k(P_h(x, x'), x'')$ are $a_i w_i^{-1}$,$a'_j(v'_j)^{-1}$ for some word w_i,v_j representing $h[a_i]$, $k[a'_j]$ respectively. Substituting v'_j in place of a'_j in w_i we get a word u_i in $a''_1, \dots, a''_l$ representing $kh[a_i]$. The substitution corresponds to an isotopy in $\partial D_n(\alpha) \sharp X_1(\alpha', \alpha'')$ which passes over 2-handles attached by $a'_j(v'_j)^{-1}$. After this isotopy we get attaching maps $a_i u_i^{-1}$, $a'_j(v'_j)^{-1}$ which are those of surgeries leading to $P_{kh}(x, P_k(x', x''))$. The same argument applies to 3-handles.

The uniqueness lemma 2.3 says that there is a unique class $\theta \in P_n(\alpha)$ containing metrics which extend to $D_n(\alpha)$. The class is the unit of $P_n(\alpha)$.

5.3. LEMMA. $P_f(\theta, x) = x$.

Proof. Let W be the cobordism of Lemma 5.1. The sum

$$Y = D_n(\alpha) \cup_{\partial D_n(\alpha) \setminus int(D^{n-1})} W$$

is diffeomorphic to the product $\partial D_n(\alpha') \times I$ by the argument we used in 5.1 to get the diffeomorphism $X_2 = D_n(\alpha')$. Let the metric $g^* \in PSC(W)$ extend $\theta \sharp g$. By the definition of θ there is an extension of g^* to Y and this is the required concordance.

Let us denote the identity homomorphism by 1.

5.4. LEMMA. $P_1(x,y) = P_1(y,x)$.

Proof. Let $x = [g]$, $y = [g']$, and $\{W_1, g_1\}$ (resp. $\{W_2, g_2\}$) denote the positive scalar curvature cobordism in the definition of $P_1(x,y)$ (resp. in the definition of $P_1(y,x)$). Then $W_1 \cong W_2 = W$ and gluing together the copies of $\partial D_n(\alpha) \sharp \partial D_n(\alpha)$ in W_1 and W_2 by the diffeomorphism changing the summands we get a positve scalar curvature metric on $W \cup (-W)$. The double is cobordant relative to the boundary $\partial D_n(\alpha) \cup \partial D_n(\alpha)$ to the product $\partial D_n(\alpha) \times I$ by a cobordism with handles of index 3 and 4. Now 2.1 gives a concordance we need.

For every group π let $i : \pi \longrightarrow \pi$ be the antihomomorphism $x \longrightarrow x^{-1}$.

5.5. LEMMA. $P_i(x,x) = \theta$.

Proof. Consider a 1-handle $h = D^1 \times D^{n-1}$ in $D_n(\alpha)$. Addition of a 2-handle with the attaching map aa' to $\partial D_n(\alpha) \sharp \partial D_n(\alpha)$ where a, a' correspond to h in the two copies of $\partial D_n(\alpha)$ is the same as forming the product $D^1 \times D^{n-1} \times I$ and taking the boundary. The metrics we have now on the boundary have equal restrictions to $D^1 \times S^{n-1} \times \{0\}$ and to $D^1 \times S^{n-1} \times \{1\}$. Therefore the metric extends from the boundary to $D^1 \times D^{n-1} \times I$, i.e. extends to $D_n(\alpha)$. A similar argument applies to 2-handles.

Define a juxtaposition in $P_n(\alpha)$ by

$$x + y = P_1(x,y).$$

5.6. PROPOSITION. $\{P_n(\alpha), +\}$ is an abelian group for $n > 5$.

Proof. Take θ as the unit and define the inverse by $-x = P_i(x,\theta)$. The proposition follows from 5.2 -5.5.

For a homomorphism $f : \pi \longrightarrow \pi'$ and arbitrary finite presentations α, α' of π and π' respectively, define

$$f_*(x) = P_f(x,\theta).$$

From 5.2. we readily get that $f_* : P_n(\alpha) \longrightarrow P_n(\alpha')$ is a homomorphism:

$$\begin{aligned} f_*(x) + f_*(y) &= P_1(P_f(x,\theta), P_f(y,\theta)) = P_f(x, P_1(\theta, P_f(y,\theta))) \\ &= P_f(x, P_f(y,\theta)) = P_f(P_1(x,y),\theta) = f_*(x+y). \end{aligned}$$

Similarly, $(gf)_* = g_* f_*$ and $id_* = id$.

Denote by $\mathcal{G}$ the category whose objects are pairs $\{\pi, \alpha\}$, where π is a group and α is a finite presentation of π, and morphisms are group homomorphisms. The following is the main result of this section.

5.7. THEOREM. P_n is a covariant functor from $\mathcal{G}$ to the category of abelian groups for any $n > 5$.

Remark. Presentations play a role analogous to the role of triangulations in simplicial homology theory, and for any two presentations α, β of a group π the isomorphism $P_1 : P_n(\alpha) \longrightarrow P_n(\beta)$ yields a canonical identification of $P_n(\alpha)$ with $P_n(\beta)$.

As in the case of homology theories, we have the *reduced functor* $\tilde{P}_*$ defined as the kernel of the homomorphism $P_*(\pi) \longrightarrow P_*(1)$ induced by the homomorphism $\pi \longrightarrow 1$.

Remark. Let $B_n(\pi,\alpha) \subseteq P_n(\alpha)$ denote the subgroup consisting of these metrics in $PSC(\partial D_n(\alpha))$ which extend to a compact spin manifold M with boundary $\partial M = \partial D_n(\alpha)$. B_n is functorial and it was considered by Jonathan Rosenberg [R]. Known invariants of positive scalar curvature metrics are related to B_n, thus only B_n can be computed in some cases. It is not known even whether $B_n \neq P_n$.

5.8. Remark. The homomorphism $P_n(\pi,\alpha) \longrightarrow P_n(1)$ has a simple geometric description. If $[g] \in P_n(\pi)$, then its image is given by the metric on S^{n-1} defined as follows. Add to $\partial D_n(\alpha)$ handles of index 2 along generators of $\pi_1 \partial D_n(\alpha) \cong \pi$, then add 3-handles to kill π_2. It is easy to check that the result of these surgeries is S^{n-1}, and the image of $[g]$ is the metric obtained by extending g to a positive scalar curvature metric on the trace of the surgeries. A similar description applies to the projection $\pi * \pi' \longrightarrow \pi' : g_1 g_1' \dots g_k g_k' \longrightarrow g_1' \dots g_k'$, where $*$ denotes the free product of groups.

6. Bordism interpretation of P_n.

Consider a pair of topological spaces (X,A). We shall define geometric bordism groups $\Omega_n^{rel}(X,A)$ which give an alternative description of the exact sequence 5.1. The groups consist of bordism classes of quadruples (M,V,g,ϕ), where M is a compact spin manifold of dimension n, V is a codimension 0 compact submanifold of ∂M g is a PSC metric on V which is product near the boundary, $\phi : (M, \partial M \setminus int(V)) \longrightarrow (X,A)$ is a continuous map. The cobordism relation is given by relative cobordism $(N,W,\bar{g},\bar{\phi})$, with the same assumptions as for objects. The juxtaposition is induced by the disjoint sum and as always it can be replaced by the connected sum. It is easy to see that $\Omega_n^{rel}(X,A)$ are abelian groups and that the following sequence

6.1. $\dots \longrightarrow \Omega_n^{Spin}(X,A) \longrightarrow \Omega_n^{rel}(X,A) \longrightarrow \Omega_n^{PSC}(X,A) \longrightarrow \Omega_n^{Spin}(X,A) \longrightarrow \dots$

is exact for any pair (X,A).

Let G be a group with a finite presentation α. Then for $n > 5$ the map

$$\Phi_\alpha : P_n(G,\alpha) \longrightarrow \Omega_n^{rel}(BG)$$

is defined by $[g] \mapsto (D_n(\alpha), \partial D_n(\alpha), g, \phi_\alpha)$ where ϕ_α is the canonical map induced by α.

6.2. THEOREM. Let $n > 5$. For every finite presentation α of G the map $\Phi_\alpha : P_n(G,\alpha) \longrightarrow \Omega_n^{rel}(BG)$ is a group isomorphism.

Proof. The juxtaposition in P_n is given by a cobordism $D_n \cup \tilde{D}_n$ to D_n extending a positive scalar curvature cobordism on ∂D_n. Since both are cobordisms in BG, thus Φ_α is a group homomorphism. If $\Phi_\alpha(g) = 0$, then there is a spin (n+1)-manifold N which contains D_n in ∂N and admits an extension of g to a PSC metric on $\partial N \setminus D_n$. One can assume also that $\pi_1 D_n \longrightarrow \pi_1 \partial N$, $\pi_1 \partial N \longrightarrow \pi_1 N$ are isomorphisms, thus N admits a PSC metric [G]. From this, using arguments of theorem 5.1. (ii) we see that g extends to $D_n(\alpha)$, hence is the zero element of P_n. The map Φ_α is onto by the standard application of the surgery theorem 2.1 to a singular manifold in BG.

7. The obstruction homomorphism.

Let π be a finitely presented group. By section 4, every class of $\Omega_n^{Spin}(B\pi)$, $n \geq 5$, contains a representative (M,f) such that $f_* : \pi_1 M \longrightarrow \pi$ is an isomorphism and

$\pi_2 M = 0$. For every finite presentation α of π we will construct an embedding (unique up to isotopy) $\phi : D_n(\alpha) \longrightarrow M$ such that
(i) ϕ preserves spin structures,
(ii) $\phi_* : \pi \longrightarrow \pi_1 M$ is the inverse of f_*,
(iii) there exists a handle decomposition of M which agrees with the decomposition of $D_n(\alpha)$ on handles of indices smaller than 3.

Define first ϕ on each S^1 in the wedge $V = S^1 \vee \ldots \vee S^1 \subseteq D_n(\alpha)$ to be an embedding in the class $f^{-1}[a_i]$, where a_i is the generator in the presentation α corresponding to this copy of S^1. The trivializations of the tangent bundle of $D_n(\alpha)$ (given by the embedding to $\mathbf{R}^n$) and of $TM|\phi(S^1)$ (given by the orientation of M) determine trivializations of the normal bundles of the circles in $D_n(\alpha)$ and in M respectively. Using these trivializations we extend ϕ to an embedding of 1-handles of $D_n(\alpha)$ such that (i) and (ii) are satisfied. Since f_* is an isomorphism, so for any attaching map $r_i : S^1 \longrightarrow V$, corresponding to a relator r_i, the composition ϕr_i is nullhomotopic. Thus ϕ extends to the rest of $D_n(\alpha)$ using the trivializations of normal bundles given by the spin structure of M. The extension is unique since $\pi_2 M = 0$ and a trivialization of a vector bundle over D^2 extends uniquely a given trivialization over S^1.

Let M_1 be the image of the sum of 1-handles and the 0-handle in $D_n(\alpha)$, $M_2 = \phi(D_n(\alpha))$. Then by the construction of ϕ, $\pi_1(M, M_2) = \pi_2(M, M_2) = 0$. Thus there is a handle decomposition of $M \setminus int(M)$ on ∂M_2 with all handles of indices greater than 2. Together with the handles inherited from $D_n(\alpha)$ we get a handle decomposition satisfying (iii).

By 2.1 there exists a metric of positve scalar curvature on $M \setminus int(M)$. Denote its restriction to ∂M_2 by g. Define $\delta(M, f) = [\phi^* g] \in P_n(\alpha)$. The following theorem is a straightforward generalization of section 2 in [H1], and we omit the proof.

7.1. THEOREM. Let π be a finitely presented group, $n > 5$.
(i) $\delta(M, f)$ depends only on the cobordism class $[M, f] \in \Omega_n^{Spin}(B\pi)$ and it defines a homomorphism $\Omega_n^{Spin} \longrightarrow P_n(\alpha)$.
(ii) $\delta x = 0$ if and only if for any (M, f) representing x and such that $f_* : \pi_1 M \longrightarrow \pi$ is an isomorphism there exists a metric of positive scalar curvature on M.
(iii) The following sequence is exact and functorial with respect to group homomorphisms

$$\ldots \longrightarrow \Omega_n^{PSC}(B\pi) \longrightarrow \Omega_n^{Spin}(B\pi) \xrightarrow{\delta} P_n(\pi, \alpha) \longrightarrow \Omega_{n-1}^{PSC}(B\pi) \longrightarrow \ldots \longrightarrow P_4(\pi, \alpha).$$

8. Additivity of P_* and groups satisfying the Gromov-Lawson conjecture.

Let π,π' be groups with finite presentations α,α'. Denote by $\alpha * \alpha'$ the presentation $(a_1, \ldots, a_s, a'_1, \ldots, a'_t; r_1, \ldots, r_u, r'_1, \ldots, r'_v)$ of the free product $\pi * \pi'$. By i, i' we denote the canonical embeddings of π, π' into $\pi * \pi'$.

8.1. Definition. Define the map $j(\alpha, \alpha') : P_n(\alpha) \times P_n(\alpha') \longrightarrow P_n(\alpha * \alpha')$ by $j(\pi, \pi')(x, x') = i(x) + i'(x') = [g \natural g']$, where $x = [g] \in P_n(\pi, \alpha)$, $x' = [g'] \in P_n(\pi', \alpha')$.

8.2. Proposition. $j(\alpha, \alpha')$ is a group isomorphism.

Proof. We have epimorphisms p, p' from $P_n(\pi * \pi')$ to $P_n(\pi)$ and $P_n(\pi')$ induced by projections from $\pi * \pi'$ to π and π' (cf. remark 5.8). The homomorphism $p \times p'$ is a left inverse to j, hence j is a monomorphism. By the theorem of Schoen and Yau (cf. [S], but no good account of this theorem is known to the author) on existence of stable minimal submanifolds, every metric on $\partial D_n(\alpha)\sharp\partial D_n(\alpha')$ can be deformed to a connected sum $g\sharp g'$ of metrics. Thus the homomorphism j is an epimorphism as well. A different method of splitting, with no use of the existence theorem, is given in [H2].

Under the identification of 8.2. we have the following additivity of δ.

8.3. Theorem. $\delta_{\pi*\pi'} = \delta_\pi \times \delta_{\pi'}$.

Proof. The equality means that $\delta_{\pi*\pi'}[M\sharp M']$ is represented by the metric $g\sharp g'$, if g, g' represent $\delta_\pi[M]$, $\delta_{\pi'}[M']$. Consider the handle decompositions of M and M' as required in the definition of δ. They give a handle decomposition of the connected sum $M\sharp M'$ with one additional handle to connect the 0-handles in M, M' and one n-1 handle which connects n-handles of the decompositions. As the result we get a decomposition corresponding to the presentation $\alpha * \alpha'$. Then proceed as in the definition of δ. At the boundary of $(M\sharp M')_{n-3} = M_{n-3}\sharp M'_{n-3}$ we get $g\sharp g'$, since the old handles do not touch the connecting 1-handle. Since δ depends only on the bordism class, the proof is complete.

8.4. Definition. We say that π satisfies the rational Gromov-Lawson conjecture if for every closed spin manifold M with fundamental group $\pi_1 M \cong \pi$, $\hat{a}_{\mathbf{Q}}(M) = 0$ if and only if the connected sum of some finite number of copies of M admits a metric with positive scalar curvature.

8.5. Theorem. The class of groups satisfying the rational Gromov-Lawson conjecture is closed with respect to the free product.

Proof. Let b_π denote the homomorphism $\delta(\pi)\otimes\mathbf{Q}$. Since vanishing of $\hat{a}_{\mathbf{Q}}$ implies existence of a metric with positive scalar curvature on $M\sharp\ldots\sharp M$ for every group π (cf. Introduction), thus $ker(\hat{a}_{\mathbf{Q}}(\pi)) \subseteq ker(b_\pi) = \{[M] : \pi_1 M \cong \pi, M \in PSC\}$. The inclusion yields a natural (with respect to group homomorphisms) transformation η_π : $KO_*(B\pi)\otimes\mathbf{Q} \longrightarrow P_*\otimes\mathbf{Q}$ making commutative the diagram

$$\begin{array}{ccc} \tilde{\Omega}_n^{Spin}(B\pi)\otimes\mathbf{Q} & \xrightarrow{\hat{a}} & P_n(\pi)\otimes\mathbf{Q} \\ & \searrow \quad\quad \eta \nearrow & \\ & \tilde{KO}_n(B\pi)\otimes\mathbf{Q}. & \end{array}$$

Now, if $\pi = \pi_1 * \pi_2$, then by theorem 8.3 the diagram splits to the product of corresponding diagrams for π_1 and π_2, hence the theorem follows.

References.

[CM] A. Connes, H. Moscovici: Cyclic cohomology, the Novikov conjecture and hyperbolic groups. Topology 29, 345-388 (1990).

[G] P. Gajer: Riemannian metrics of positive scalar curvature on compact manifolds with boundary. Ann. Global Anal. Geom. 5, 179-191 (1987).

[GL1] M. Gromov, B. Lawson: The classification of simply connected manifolds of positive scalar curvature. Ann. Math. 111, 423-434 (1980).

[GL2] M. Gromov, B. Lawson: Positive scalar curvature and the Dirac operator on complete Riemannian manifolds. Publ. IHES 58, 295-408 (1983).

[H1] B. Hajduk: Metrics of positve scalar curvature on spheres and the Gromov-Lawson conjecture. Math. Ann. 208, 409-415 (1988).

[H2] B. Hajduk: Splitting metrics with positive scalar curvature along submanifolds. (to appear)

[Ha] M. Hall: The theory of groups. Macmillan. New York 1959.

[Hi] N. Hitchin: Harmonic spinors. Adv. Math. 14, 1-55 (1974).

[K] M. Kreck: Positive scalar curvature, $P_2(H)$ and elliptic homology. 29. Arbeitstagung Bonn, 23. - 29. Juni 1990. Preprint Max-Planck-Institut.

[KS] S. Kwasik, R. Schulz: Positive scalar curvature and periodic fundamental groups. Comment. Math. Helv. 65, 271-286 (1990).

[L] A. Lichnerowicz: Spineurs harmoniques. C.R. Acad. Sci. Paris, 257, 7-9 (1963).

[Mi1] J. Milnor: Lectures on Morse theory. Princeton.

[Mi2] J. Milnor: Remarks concerning spin manifolds. A Symposium in Honor of Marston Morse, 55-62. Princeton University Press 1965.

[M] T. Miyazaki: On the existence of positive scalar curvature metrics on non-simply connected manifolds. J. Fac. Sci. Tokyo 30, 549-561 (1984).

[R1] J. Rosenberg: C^*-algebras, positve scalar curvature, and the Novikov conjecture II, Proc. U.S.-Japan Seminar on Geometric Methods in Operator Algebras, Kyoto 1963.

[R2] J. Rosenberg: C^*-algebras, positive scalar curvature, and the Novikov conjecture III, Topology 25, 319-336 (1986).

[R3] J. Rosenberg: The KO-assembly map and positive scalar curvature (Preprint, December 1989).

[S] R.M. Schoen: Minimal surfaces and positive scalar curvature. Proceedings of ICM Warszawa 1983, 575-578.

[SY] R. Schoen, S.T. Yau: On the structure of manifolds with positive scalar curvature. Manuscr. Math. 28, 159-183 (1979).

This paper is in final form and no version will appear elsewhere.

INSTITUTE OF MATHEMATICS
WROCŁAW UNIVERSITY
pl. Grunwaldzki 2/4
50-384 Wrocław, Poland

Compact manifolds with 1/4–pinched negative curvature

Ursula Hamenstädt

Let M be a compact n–dim. Riemannian manifold of negative sectional curvature K_M, normalized in such a way that the maximum of K_M equals -1. The geodesic flow ϕ^t $(t \in \mathbb{R})$ acts on the unit tangent bundle T^1M of M; it is a dynamical system generated by a smooth vector field X on T^1M, the so called geodesic spray. There are ϕ^t–invariant continuous foliations W^{su}, W^{ss} on T^1M which are called the strong unstable resp. the strong stable foliation. Each leaf of W^i is an $(n-1)$–dim. C^∞–immersed submanifold of T^1M. The tangent bundles of these leaves define a Hölder–continuous ϕ^t–invariant subbundle TW^i of TT^1M $(i = su,ss)$ in such a way that $E = TW^{su} \oplus TW^{ss}$ is a smooth subbundle of TT^1M whose orthogonal complement equals the 1–dim. subbundle T^0 which is spanned by the geodesic spray; thus we have $TT^1M = T^0 \oplus E$.

Let $Gr(n-1)$ be the smooth fibre bundle over T^1M whose fibre at v consists of the Grassmannian of $(n-1)$–dim. subspaces of E_v. The bundles TW^{su}, TW^{ss} then define Hölder–continuous sections σ^{su}, σ^{ss} of $Gr(n-1)$. In general these sections are not smooth (compare [K], [F–K], [F], [F–L]), but if the curvature of M is strictly 1/4–pinched they are of class C^1 ([H–P]).

In this note we investigate the regularity of σ^{su}, σ^{ss} under the assumption that the curvature of M is 1/4–pinched, but not necessarily strictly. For this we call a section σ of $Gr(n-1)$ almost everywhere differentiable if there is an open, ϕ^t–invariant subset V of T^1M such that for every ϕ^t–invariant ergodic Borel–probability measure ν on T^1M with $\nu(V) = 1$ the section σ^{su} is differentiable at ν–almost every $v \in T^1M$, with ν–measurable differential.

Our purpose is to show:

Theorem: If the curvature of M is 1/4–pinched then σ^{su} and σ^{ss} are almost everywhere differentiable.

Assume that M is such that its curvature is contained in $[-4,-1]$, with curvature maximum -1, and that M is not locally symmetric. Recall from [H1] the notion of the hyperbolic rank h–rank (M) of M. We showed in [H1] that this rank is nonzero only for locally symmetric spaces; thus we may assume that h–rank $(M) = 0$. Denote by $\langle\,,\,\rangle^i$ the Riemannian metric on the leaves of W^i which is lifted from the Riemannian metric on M and let $\|\ \|^i$ be the associated norm ($i = su, u, s, ss$). Then $\|d\phi^t X\|^{su} \geq e^t\|X\|^{su}$ for all $X \in TW^{su}$, all $t \geq 0$, moreover by 3.2 of [H1] there is a nonempty open set $\Omega \subset T^1M$ and numbers $\tau > 0$, $\epsilon > 0$ such that $\|d\phi^\tau X\|^{su} \geq e^{\tau(1+2\epsilon)}\|X\|^{su}$ for all $v \in \Omega$ and all $X \in T_vW^{su}$.

Proposition 1: Let ν be a flow–invariant ergodic Borel–probability measure on T^1M which is positive on open sets. Then the smallest positive Lyapunov exponent of ϕ^t with respect to ν is strictly larger than 1.

Proof: Let $\Omega \subset T^1M$, $\tau > 0$ be as above, let χ be the characteristic function of Ω and let ν be a ϕ^t–invariant ergodic Borel–probability measure on T^1M which is positive on open sets. Then $c = \nu(\Omega) > 0$ and by the Birkhoff ergodic theorem we have $\lim_{k\to\infty}\frac{1}{k}\sum_{i=0}^{k-1}\chi(\phi^{i\tau}w) = c$ for ν–almost every $w \in T^1M$. Let $w \in T^1M$ be such a point and let $k_0 > 0$ be such that $\sum_{i=0}^{k-1}\chi(\phi^{i\tau}w) \geq ck/2$ for all $k \geq k_0$. Then $\|d\phi^{k\tau}X\|^{su} \geq e^{k\tau(1+\epsilon c)}\|X\|^{su}$ for all $k \geq k_0$ and all $0 \neq X \in T_wW^{su}$ and consequently the smallest Lyapunov exponent of ϕ^t with respect to ν is not smaller than $1+\epsilon c > 1$ as claimed.

With the notations of the beginning of this sections let $P^i : E \longrightarrow TW^i$ ($i = su, ss$) be the canonical projections and define a Finsler metric $|\ \ |$ on E by $|X| = \max\{\|P^{su}X\|^{su}, \|P^{ss}X\|^{ss}$. The curvature assumption $k \geq -4$ implies that with respect to this Finsler metric on E the operator norm of $d\phi^\tau$ does not exceed $e^{2\tau}$ for $\tau \geq 0$. If $E = E^{su} \oplus E^{ss}$ is anydecomposition of E into continuous (n–1)–dim. subbundles then $|\ \ |$ introduces a Finsler metric on the bundle $(E^{su})^* \otimes E^{ss}$ by assigning to $L \in (E^{su})^*_v \otimes E^{ss}_v$, viewed as a linear mapping of E^{su}_v into E^{ss}_v, its operator norm with respect to $|\ \ |$. We denote this Finsler metric again by $|\ \ |$.

Recall that for $v \in T^1M$ the vector space $(T_vW^{su})^* \otimes T_vW^{ss}$ has a natural identification with the open subset $Q_v = \{H \in Gr(n-1)_v \mid P^{su}H = T_vW^{su}\}$ of $Gr(n-1)_v$. The set $Q = \bigcup_v Q_v$ is an open neighborhood of $\sigma^{su}(T^1M)$ in $Gr(n-1)$, diffeomorphic to $(TW^{su})^* \otimes TW^{ss}$. In this identification the action of ϕ^t on Q is given by the <u>graph transform</u> $d\phi^\tau_\#$ of ϕ^t which is defined by $(d\phi^\tau)_\#$ (graph L) = graph$(d\phi^\tau)_\# L)$. For every $v \in T^1M$ the graph transform $(d\phi^\tau)_\#$ maps the ball of radius r in $((TW^{su})^* \otimes TW^{ss})_v$ into the ball of radius $e^{-2\tau}r$ in $((TW^{su})^* \otimes TW^{ss})_{\phi^\tau v}$ (see [H–P] p. 237).

Let $U \subset \Omega$ be a nonempty open set whose closure $\bar{U}$ in T^1M is contained in Ω and let $\beta \in (0,1/2]$. Then there is a smooth approximation E^i of the bundles TW^i $(i = su, ss)$ with the following properties (see [H–P]):

i) $E = E^{su} \oplus E^{ss}$.

ii) The disk bundle $G = \{L \in (E^{su})^* \otimes E^{ss} \mid |L| \leq 1\}$ is naturally isomorphic to a closed subset of Q and is left invariant under the graph transform $f = d\phi^\tau_\#$ of ϕ^τ. The action of f on the space Σ of continuous sections of G via $f_\#\sigma(v) = f\circ\sigma\circ\phi^{-\tau}(v)$ $(v \in T^1M)$ has a unique continuous fixed point σ_f corresponding to σ^{su}. If $\sigma \in \Sigma$ is arbitrary then $f^k_\#\sigma \longrightarrow \sigma_f$ $(k \longrightarrow \infty)$ uniformly on T^1M.

iii) For $\alpha = \alpha(\beta) = \beta\tau/(1-\beta)$ the Lipschitz constant of the restriction of f to every fibre G_w does not exceed $e^{-2\tau+\alpha\epsilon}$; if $w \in \Omega$ then this Lipschitz constant is not larger than $e^{-2(1+\epsilon)\tau}$.

Following [H–P–S] we may assume that the bundle $L(E) = (E^{su})^* \otimes E^{ss}$ over T^1M is trivial. Thus every section $\sigma : T^1M \longrightarrow L(E)$ of $L(E)$ is of the form $\sigma(X) = (X,s(X))$ for a map s of T^1M into the typical fibre Z of $L(E)$. Let d be the distance on T^1M induced by the Riemannian metric and define slope$_x(\sigma)$ of σ at x to be the local dilation of s at x with respect to the metric $| \ |_x$ on Z induced from the Finsler structure of $L(E)_x$, i.e. $\text{slope}_x(\sigma) = \limsup_{u \to x} \frac{|s(u)-s(x)|_x}{d(x,u)}$. Recall that for $v \in T^1M$ and $z \in G_v$ we have $T_zL(E) = T_v(T^1M) \oplus L(E)_v$; for $\ell > 0$ we then define $\text{cone}_z(\ell) = \{X+Y \mid X \in T_vT^1M, Y \in L(E)_v, |Y| \leq \ell|X|\}$.

Lemma 2: Let $v \in T^1M$, $z \in G_v$ and for $i > 0$ define $\kappa_i > 0$ to be the Lipschitz constant of the restriction of $d_{f^i z} f$ to $L(E)_{\phi^{\tau i} v}$. Let $\chi(0) = 1$ and for $k > 0$ define inductively $\chi(k) = 1 + \chi(k-1)\kappa_{k-1} e^{2\tau}$; there is a number $\ell(\beta) > 0$ such that $d_z f^k(\mathrm{cone}_z(\ell)) \subset \mathrm{cone}_{f^k z}(\chi(k)\ell)$ for all $k \geq 0$, all $\ell \geq \ell(\beta)$.

Proof: Since by our assumption the bundle $L(E)$ is trivial the differential $d_z f$ of f at z can be expressed as $d_z f = \begin{bmatrix} A_z & 0 \\ C_z & B_z \end{bmatrix}$ where $A_z : T_v T^1M \longrightarrow T_{\phi^\tau v} T^1M$, $C_z : T_v T^1M \longrightarrow L(E)_{\phi^\tau v}$ and $B_z : L(E)_v \longrightarrow L(E)_{\phi^\tau v}$. Let $c = \sup\{\|C_z\| \mid z \in G\}$ and define $\ell(\beta) = ce^{2\tau}$.

We show the lemma for $\ell > \ell(\beta)$ by induction on k. The case $k = 0$ being trivial, assume that the lemma is true for $k-1$. Let $X+Y \in \mathrm{cone}_{f^{k-1} z}(\chi(k-1)\ell)$ and let $(d_{f^{k-1} z} f)(X+Y) = X' + Y'$. Then $|Y'| \leq \kappa_{k-1}|Y| + c|X| \leq \kappa_{k-1}\chi(k-1)\ell\, e^{2\tau}|X'| + e^{2\tau}c|X'| \leq (\kappa_{k-1}e^{2\tau}\chi(k-1)+1)\ell|X'| = \chi(k)\ell|X'|$ which shows the lemma. ∎

For $w \in T^1M$ and every integer $k > 0$ let $\#(w,k)$ be the cardinality of the set $\{\phi^{-i\tau}w \mid 1 \leq i \leq k\} \cap U$. For $j > 0$ define $A(\beta,j) = \{v \in T^1M \mid \#(v,k) \geq k\beta$ for all $k \geq j\}$. For $v \in T^1M$ and $\rho > 0$ let moreover $B^{su}(v,\rho)$ be the ball of radius ρ about v in $W^{su}(v)$ with respect to the distance d^{su} induced by the Riemannian metric $\langle\, , \rangle^{su}$. Assume that $r > 0$ is sufficiently small that $B^{su}(v,r)$ is diffeomorphic to a ball in $\mathbb{R}^{n-1}$ for all $v \in T^1M$.

Lemma 3: Let $\beta \in (0,1/2]$ and $E = E^{su} \oplus E^{ss}$ be as before. Then for every $j > 0$ there is a number $\ell(\beta,j) > 0$ such that $\mathrm{slope}_w(\sigma^{su}) \leq \ell(\beta,j)$ for all $w \in \cup\{B^{su}(v,r) \mid v \in A(\beta,j)\}$.

Proof. Choose $\ell \geq \ell(\beta)$ in such a way that there is a smooth section $\sigma : T^1M \longrightarrow G$ with $\mathrm{slope}_x\sigma \leq \ell$ for all $x \in T^1M$. Since $f_\#^k\sigma \longrightarrow \sigma^{su}$ uniformly on T^1M we only have to show that for every $j > 0$ there is a number $\rho(j) > 0$ such that $\mathrm{slope}_w(f_\#^k\sigma) \leq \rho(j)\ell$ for all $v \in A(\beta,j)$ and all $w \in B^{su}(v,r)$ (compare [H–P–S]).

For $w \in T^1M$ and $k > 0$ define $\omega(w,k)$ to be the cardinality of the set $\{\phi^{-i\tau}w \mid 1 \le i \le k\} \cap \Omega$. Since $\overline{U} \subset \Omega$, i.e. $d(\overline{U},T^1M-\Omega) > 0$ and $d(\phi^{-t}v,\phi^{-t}w) \longrightarrow 0$ $(t \longrightarrow \infty)$ uniformly in $v \in T^1M$ and $w \in B^{su}(v,r)$ there is a number $R > 0$ such that $\omega(w,k) \ge \#(v,k)-R$ for all $v \in T^1M$ and all $w \in B^{su}(v,r)$.

Let $v \in A(\beta,j)$, $w \in B^{su}(v,r)$, $k \ge j$ and $z = \sigma(\phi^{-k\tau}v)$. For $i \ge 0$ denote by κ_i the Lipschitz constant of the restriction of $d_{f^i_z} f$ to $L(E)_{\phi^{-k\tau+i\tau}w}$ and as in lemma 3 let $\chi(0) = 1$ and $\chi(k) = 1 + \chi(k-1)\kappa_{k-1}e^{2\tau}$ $(k \ge 1)$. Now by iii) above we have $\kappa_{k-i} \le e^{-2\tau(1+\epsilon)}$ if $\phi^{-i\tau}w \in \Omega$ and $\kappa_{k-i} \le e^{-2\tau+\alpha\epsilon}$ otherwise; thus for $k \ge m \ge j$ the coefficient of $e^{2m\tau}$ in the expression $\chi(k) = 1 + e^{2\tau}\kappa_{k-1} + e^{4\tau}\kappa_{k-1}\kappa_{k-2} + \dots + e^{2k\tau}(\kappa_{k-1}\kappa_{k-2}\cdots\kappa_0)$ does not exceed $e^{\alpha\epsilon(m-m\beta+R)}e^{-2\tau\epsilon(\beta m-R)} = e^{R\epsilon(2\tau+\alpha)}e^{-\beta\tau\epsilon m}$. On the other hand, for $m < j$ this coefficient is less or equal than $e^{\alpha\epsilon m}$ by iii). Together this yields $\chi(k) \le \sum_{m=0}^{j-1} e^{\alpha\epsilon m} + e^{R\epsilon(2\tau+\alpha)}\left(\sum_{m=1}^{\infty} e^{-\beta\tau m\epsilon}\right) = \rho(j) < \infty$. Now lemma 3 shows $\mathrm{slope}_{f^k_z}(f^k_{\#}\sigma) \le \rho(j)\ell$ for all $k \ge 0$ whence the lemma. ∎

Now we are ready for the proof of our theorem. Define $V = \bigcup_{t\in\mathbb{R}} \phi^t\Omega$; then V is an open, ϕ^t-invariant subset of T^1M. Let ν be a ϕ^t-invariant ergodic Borel-probability measure on T^1M with $\nu(V) = 1$. Since $V = \bigcup\{\phi^t\Omega \mid t \text{ is rational}\}$ and $\nu(v) = 1$ we necessarily have $\nu(\Omega) > 0$. The arguments in the proof of proposition 1 then show that there is $\beta > 0$ such that $\nu(\bigcup_{j>0} A(\beta,j)) = 1$. Now by lemma 4 and the arguments of [H-P-S] (p. 33-35) the section σ^{su} is differentiable on $A(\beta,j)$ for every $j > 0$, with continuous differential. Moreover the differential of σ^{su} on $A_\beta = \bigcup_{j>0} A(\beta,j)$ is the limit as $k \longrightarrow \infty$ of the differentials of the sections $f^k_{\#}\sigma$ for any fixed smooth element of Σ, i.e. the differential of σ^{su} is indeed ν-measurable.

References

[F] R. Feres, Geodesic flows on manifolds of negative curvature with smooth horospheric foliations, Thesis, Caltech 1989.

[F–K] R. Feres, A. Katok, Invariant tensor fields of dynamical systems with pinched Lyapunov exponents and rigidity of geodesic flows. Erg. Th. & Dyn. Sys. 9 (1989), 427–432.

[F–L] P. Foulon, F. Labourie, Flots d'Anosov à distributions de Liapounov différentiables, preprint.

[H1] U. Hamenstädt, A geometric characterization of compact locally symmetric spaces, to appear in J. Diff. Geo.

[H2] U. Hamenstädt, Metric and topological entropies of geodesic flows, to appear in Ann. Math.

[H–P] M. Hirsch, C. Pugh, Smoothness of horocycle foliations, J. Diff. Geom. (1975), 225–238.

[H–P–S] M. Hirsch, C. Pugh, M. Shub, Invariant manifolds, Lecture Notes in Math. 583, Springer 1977.

[K] M. Kanai, Geodesic flows on negatively curves manifolds with smooth stable and unstable foliations, Erg. Th. & Dyn. Sys. 8 (1988), 215–239.

Max–Planck–Institut für Mathematik
Gottfried–Claren–Straße 26
5300 Bonn 3
Germany

The Geometry of Moduli Spaces of Stable Vector Bundles over Riemann Surfaces

Jürgen Jost and Xiao-Wei Peng

§ 1 Introduction

In this paper, we study the geometry of moduli spaces of stable vector bundles over Riemann surfaces. The natural metric on such moduli spaces is induced by an L^2-metric on certain spaces of connections. We relate this metric to a variational problem for bundle maps. This gives us a convenient method for computing e.g. the curvature tensor of the metric on the moduli space, thereby reproducing a formula of Zograf-Takhtadzhyan [14]. We then turn to studying the effect of variations of the underlying Riemann surface on the geometry of the corresponding moduli spaces of stable bundles, using the same variational method. The formulae we obtain seem to be of interest in connection with recent results of Axelrod-della Pietra-Witten [1], Hitchin [4], and Beilinson-Kazhdan [2]. Those authors construct a projectively flat connection on a certain bundle over the moduli space of Riemann surfaces the fiber of which is given by the space of holomorphic sections of a certain line bundle over the moduli space of stable bundles on the given surface.

We now describe the setting of the problem. Let Σ be a compact Riemann surface of genus $p > 1$. As in the work of Narasimhan-Seshadri [10], we represent Σ as H/Γ, where H is the upper half plane equipped with its hyperbolic metric, and Γ is a discrete group of automorphisms of H with a unique fix point $z_0 \in H$, i.e. $\gamma_0 z_0 = z_0$ for a unique $\gamma_0 \in \Gamma, \gamma_0 \neq \mathrm{id}$. Let $N(n,k)$ be the moduli space of stable vector bundles of rank n and degree k over Σ. By the theorem of Narasimhan-Seshadri [10], the moduli space $N(n,k)$ is isomorphic to the space of classes of irreducible representations $\rho : \Gamma \to U(n)$, where $U(n)$ is the unitary group, with normalization

$$\rho(\gamma_0) = \exp(-\frac{2\pi i k}{n})E,$$

where E is the unit matrix. Of course, ρ is only determined up to conjugation in $U(n)$.

For an irreducible representation $\rho : \Gamma \to U(n)$, the corresponding stable vector bundle E_ρ over Σ is the quotient space of $H \times \mathbb{C}^n$ by the action of Γ which is defined by the formula

$$(\gamma, z, w) \mapsto (\gamma z, \rho(\gamma) w) \quad \text{for } \gamma \in \Gamma, z \in H, w \in \mathbb{C}^n.$$

It is wellknown that $N(n,k)$ is a complex manifold of dimension $n^2(p-1)+1$ over $\mathbb{C}$. It is compact if and only if n and k are relatively prime. Otherwise, it can be compactified by adding semistable bundles, but this process creates certain singularities. As a consequence of the theory of Kodaira-Spencer, the tangent space $T_\rho N(n,k)$ of $N(n,k)$ at the point corresponding to the representation $\rho : \Gamma \to U(n)$ can be naturally identified with the Dolbeault cohomology group $H^{0,1}(\Sigma, \text{ End } E_\rho)$. There is a natural L^2-metric (Riemannian metric) for $N(n,k)$ defined by the following formula:

$$\langle \mu, \nu \rangle := -\frac{i}{2} \int_\Sigma tr(\mu \wedge \bar{\nu}^t), \text{ for } \mu, \nu \in T_\rho N(n,k) = H^{0,1}(\Sigma, \text{ End } E_\rho).$$

In case $n = 1$, of course, all this reduces to the classical theory of the Jacobian or Picard varieties of a Riemann surface.

In § 2, we shall introduce a variational integral whose solutions yield a local coordinate chart for $N(n,k)$, and whose second derivatives give the above L^2-metric. Computing derivatives of the metric with the help of this variational integral yields an easy proof of the following curvature formula of Zograf-Takhtadzhyan (cf. [12], [14]).

Theorem: The curvature tensor of the L^2-metric of $N(n,k)$ at ρ

$$\begin{aligned} R_{\mu\bar{\nu}\alpha\bar{\beta}} &= -\frac{i}{4} \int_\Sigma tr([\bar{\beta}^t, \mu] \wedge \Delta_0^{-1}[\alpha, \bar{\nu}^t]) \\ &\quad - \frac{i}{4} \int_\Sigma tr([\bar{\nu}^t, \mu] \wedge \Delta_0^{-1}[\alpha, \bar{\beta}^t]) \end{aligned}$$

where $\Delta_0 = \partial\bar{\partial}$.

In §3 we want to study the effect of variations of Σ. We consider the fiber space N_p over Teichmüller space T_p of Riemann surfaces of genus $p \geq 2$, where the fiber over $\tau \in T_p$ is the moduli space of stable vector bundles of rank n and degree k over the Riemann surface Σ with conformal structure determined by τ. We compute derivatives of our variational integral in its dependence on τ, representing tangent vectors of T_p by harmonic Beltrami differentials. First derivatives yield a generalization of the Rauch variational formula. The original formula of Rauch [11] expresses the variation of the period mapping of a Riemann surface in its dependence on $\tau \in T_p$, and it is well known that this can be considered as a formula for the variation of the Picard variety of a Riemann surface, which corresponds to the case $n = 1$ of our present setting.

Finally, our method also allows to compute second derivatives in the Teichmüller directions. These will be obtained at the end of §3.

The authors gratefully acknowledge financial support from DFG.

§ 2 The geometry of moduli spaces $N(n,k)$.

Let $E_\rho \in N(n,k)$ be induced by a class of irreducible representations

$$\rho : \Gamma \to U(n).$$

We recall the normalization $\rho(\gamma_0) = \exp(-\frac{2\pi i k}{n})E$, for a fixed $\gamma_0 \in \Gamma$.

We now want to introduce a variational functional and construct local coordinate charts for $N(n,k)$. Let $E_\rho, E_\sigma \in N(n,k)$ be defined by irreducible representations ρ and σ, resp. An isomorphism between the vector bundles E_ρ and E_σ over Σ is defined by a map

$$h : H \to Gl(n,\mathbb{C})$$

with transformation behaviour

$$h(\gamma z) = \sigma(\gamma)h(z)\rho(\gamma)^{-1} \quad \text{for all } z \in H, \gamma \in \Gamma.$$

For uniqueness reasons, we require a normalization, namely

$$h(z_0) = E$$

where z_0 was the unique fixed point of $\gamma_0 \in \Gamma$.

Let

$$\mathcal{M}(\rho,\sigma) = \left\{ h : H \to GL(n,\mathbb{C}) \,\Big|\, \begin{array}{l} h(\gamma z) = \sigma(\gamma)h(z)\rho(\gamma)^{-1} \\ h(z_0) = E \end{array} \right\}.$$

If $\rho = \sigma, \mathcal{M}(\rho,\rho)$ is the space of all normalized sections of the bundle $\operatorname{Aut} E_\rho$ over Σ, $\Gamma_0(\Sigma, \text{ Aut } E_\rho)$. The tangent space of $\mathcal{M}(\rho,\rho)$ at each point can be identified with $\Gamma_0(\Sigma, \text{ End } E_\rho)$, where the subscript now denotes the normalization $\eta(z_0) = 0$.

We now define an energy functional on $\mathcal{M}(\rho,\sigma)$:

$$E(h) := -\frac{i}{2}\int_\Sigma tr(h^{-1}\bar{\partial}h \wedge \overline{(h^{-1}\bar{\partial}h)}^t).$$

The Euler-Lagrange equation of this energy functional is

$$\partial(h^{-1}\bar{\partial}h) = 0, \text{ i.e. } h^{-1}\bar{\partial}h \in H^{0,1}(\Sigma, \text{ End } E_\rho), \tag{1}$$

where $H^{0,1}(\Sigma, \text{ End } E_\rho)$ is the space of all anti-holomorphic $(0,1)$–forms with values in End E_ρ over Σ. The equation (1) is always solvable for σ near ρ, since ρ is irreducible.

We denote the solution by $h = h(\rho,\sigma)$.
The solvability of (1) for σ in a neighborhood of ρ can be seen as follows:

We consider the map

$$F : \mathcal{M}(\rho,\sigma) \to \Omega^{1,1}(\Sigma, \text{ End } E_\rho)$$
$$h \mapsto \partial(h^{-1}\bar{\partial}h).$$

The linearization of the map F at $\sigma = \rho$ is given by

$$\partial\overline{\partial} : \Gamma_0(\Sigma,\ \text{End}\ E_\rho) \to \Omega^{1,1}(\Sigma,\ \text{End}\ E_\rho).$$

Since ρ is irreducible, the normalization condition implies that $\partial\overline{\partial}$ is invertible. The implicit function theorem then yields the local solvability of (1).
For a solution h of (1), $h^{-1}\overline{\partial}h$ is anti-holomorphic, since $\partial(h^{-1}\overline{\partial}h) = 0$. Using this fact, we define a map from irreducible representations to anti-holomorphic sections of $\text{End}E_\rho \otimes \wedge^{0,1}T^*\Sigma$,

$$\sigma \mapsto h^{-1}\overline{\partial}h = \mu \in H^{0,1}(\Sigma,\ \text{End}\ E_\rho).$$

In the same way as above, we see that this map is a local diffeomorphism (in a neighborhood of ρ). It consequently defines a local coordinate chart for $N(n,k)$ in a neighborhood of ρ.

Let ρ_t be a 1–parameter family of irreducible representations of Γ in $U(n)$ with $\rho_0 = \rho$, t in a neighborhood of $0 \in \mathbb{C}$ corresponding to a complex curve $t\mu$ in $H^{0,1}(\Sigma,\ \text{End}\ E_\rho)$. By the above discussion there exists a family of maps $h_t : H \to Gl(n,\mathbb{C})$, which depends smoothly on the parameter t and satisfies the equation:

$$h_t^{-1}\overline{\partial}h_t = t\mu \tag{2}$$

with

$$h_t(\gamma z) = \rho_t(\gamma)h_t(z)\rho(\gamma)^{-1}, h_t(z_0) = E \quad \text{for all } t. \tag{3}$$

Since $\rho_0 = \rho$, we have $h_0 = E$. Set

$$\dot{h}_- = \frac{\partial h}{\partial \bar{t}}, \qquad \dot{h}_+ = \frac{\partial h}{\partial t}.$$

Differentiating (2) and (3) w.r.t. t yields

$$\bar{\partial}\dot{h}_- = 0,\ \ \bar{\partial}\dot{h}_+ = \mu \tag{4}$$

and

$$\dot{h}_-(\gamma z) = \dot{\rho}_-(\gamma)\rho(\gamma)^{-1} + \rho(\gamma)\dot{h}_-(z)\rho(\gamma)^{-1} \tag{5}$$

$$\dot{h}_+(\gamma z) = \dot{\rho}_+(\gamma)\rho(\gamma)^{-1} + \rho(\gamma)\dot{h}_+(z)\rho(\gamma)^{-1}. \tag{6}$$

Putting (5) and (6) together,

$$\begin{aligned}\dot{h}_-(\gamma z) + \overline{\dot{h}(\gamma z)}^t &= (\rho_t(\gamma)\overline{\rho_t(\gamma)}^t)^\cdot + \rho(\gamma)(\dot{h}_-(z) + \overline{\dot{h}_+(z)}^t)\rho(\gamma)^{-1} \\ &= \rho(\gamma)(\dot{h}_-(z) + \overline{\dot{h}_+(z)}^t)\rho(\gamma)^{-1}.\end{aligned}$$

This means that $\dot{h}_- + \overline{\dot{h}_+}^t$ is an endomorphism of E_ρ, i.e. $\dot{h}_- + \overline{\dot{h}_+}^t \in \Gamma_0(\Sigma, \text{ End } E_\rho)$. We also have

$$\begin{aligned}\partial\bar{\partial}(\dot{h}_- + \overline{\dot{h}_+}^t) &= \partial\bar{\partial}\dot{h}_- + \bar{\partial}\partial\overline{\dot{h}_+}^t \\ &= \bar{\partial}\bar{\mu}^t \\ &= 0.\end{aligned}$$

Using partial integration over Σ we obtain

$$\int_\Sigma |\bar{\partial}(\dot{h}_- + \overline{\dot{h}_+}^t)|^2 = 0.$$

Therefore, $\dot{h}_- + \overline{\dot{h}_+}^t$ is a holomorphic section of End E_ρ over Σ.

Since E_ρ is stable, hence simple, the only holomorphic sections of End E_ρ are constant multiples of the unit matrix (see [9], [5]). Therefore

$$\partial(\dot{h}_- + \dot{h}_+^t) = 0,$$

and (4) then implies

$$\partial\dot{h}_- = -\bar{\mu}^t \quad \text{(cf. [13])}.$$

Note: Γ is not the fundamental group of Σ. The representation of Σ as H/Γ is the same as in the work of Narasimhan-Seshadri that is the projection $\pi : H \to H/\Gamma$ is ramified at most over a single point $x_0 \in H/\Gamma$ i.e. the isotropy group of Γ at $y, y \in \pi^{-1}(x_0)$, is a finite cyclic group and its order is independent of the point $y \in \pi^{-1}(x_0)$, if $y \notin \pi^{-1}(x_0)$, the isotropy group of Γ at y reduces to the identity. In case $k = 0$, we can choose Γ to be the fundamental group of Σ.

For $\rho, \sigma \in N(n, k)$, we solve

$$\begin{aligned}&h = h(\rho, \sigma) : H \to Gl(n, \mathbb{C}) \\ &\partial(h^{-1}\bar{\partial}h) = 0 \\ &h(\gamma z) = \sigma(\gamma)h(z)\rho(\gamma)^{-1}\end{aligned}$$

as above, we have

$$E(h(\rho, \sigma)) := -\frac{i}{2}\int_\Sigma tr\ (h^{-1}\bar{\partial}h \wedge \overline{(h^{-1}\bar{\partial}h)}^t)$$

and $\langle \mu, \nu \rangle = \frac{\partial}{\partial \bar{s}}\frac{\partial}{\partial t}E(h(\rho, \rho + t\mu + s\nu))|_{t=0, s=0}$, for $t, s \in \mathbb{C}$.

In order to differentiate the metric, we have to vary ρ, i.e. for $\alpha, \mu \in H^{0,1}(\Sigma, \text{ End } E_\rho)$, we have to consider

$$h : H \to Gl(n, \mathbb{C})$$

with

$$\partial(h^{-1}\bar{\partial}h) = 0 \tag{7}$$

(8) $$h(\gamma z) = \rho_{\mu+\alpha}(\gamma) h(z) \rho_\alpha(\gamma)^{-1} \quad .$$

We write

$$h = \eta g$$

with
$$\eta(\gamma z) = \rho_{\mu+\alpha}(\gamma)\eta(z)\rho(\gamma)^{-1}$$
$$g(\gamma z) = \rho(\gamma) g(z) \rho_\alpha(\gamma)^{-1}$$

where g satisfies

(9) $$\partial(g\bar{\partial}g^{-1}) = 0 \, .$$

From the above relation, i.e. $\eta := hg^{-1}$, equation (7) $\Rightarrow$

$$\frac{\partial}{\partial z}(g^{-1}\eta^{-1}(\frac{\partial \eta}{\partial \bar{z}} g + \eta \frac{\partial g}{\partial \bar{z}})) = 0 \, ,$$

and equation (9) $\Leftrightarrow$

(10) $$\eta^{-1}\frac{\partial^2 \eta}{\partial z \partial \bar{z}} - \eta^{-1}\frac{\partial \eta}{\partial z}\eta^{-1}\frac{\partial \eta}{\partial \bar{z}} + \eta^{-1}\frac{\partial \eta}{\partial \bar{z}}\frac{\partial g}{\partial z} g^{-1} - \frac{\partial g}{\partial z} g^{-1}\eta^{-1}\frac{\partial \eta}{\partial \bar{z}} = 0 \, .$$

We want to differentiate (10) w.r.t. α. We note that the α-dependence of h and hence also of η is two-fold (see (8)).
Varying the α-dependence in $\rho_{\mu+\alpha}$ will be denoted by $'$ and $[\cdot]$, and the variation of the α-dependence in $\rho_{\bar{\alpha}}^{-1}$ by $^{\cdot}$ and $\langle \cdot \rangle$. Now for $\alpha = \mu = 0$, $\eta = g = E$ (the unit matrix in $Gl\,(n, \mathbb{C})$).
Also

$$(\partial g g^{-1})^{\cdot}\langle \bar{\alpha} \rangle = \bar{\alpha}^t, \qquad (\partial g g^{-1})^{\cdot}\langle \alpha \rangle = 0$$
$$(\eta^{-1}\bar{\partial}\eta)'[\alpha] = \alpha, \qquad (\eta^{-1}\bar{\partial}\eta)'[\bar{\alpha}] = 0 \, .$$

Thus differentiating (10), using also $\partial\alpha = \alpha_z = 0$

(11)
$$\frac{\partial}{\partial z}((\eta^{-1}\frac{\partial \eta}{\partial \bar{z}})^{\cdot}\langle \alpha \rangle) = 0$$
$$\frac{\partial}{\partial z}((\eta^{-1}\frac{\partial \eta}{\partial \bar{z}})^{\cdot}\langle \bar{\alpha} \rangle) = 0 \, ,$$

thus $(\eta'^{-1}\bar{\partial}\eta)^{\cdot}\langle \alpha \rangle \in H^{0,1}(\Sigma, \text{ End } E_\rho)$.

On the other hand, $\dot{\eta}(\gamma z) = \rho(\gamma)\dot{\eta}(z)\rho(\gamma)^{-1}$ and hence $\dot{\eta}$ represents an infinitesimal endomorphism of E_ρ and $(\eta^{-1}\bar{\partial}\eta)^{\cdot}\langle \alpha \rangle$ is hence L^2-orthogonal to $H^{0,1}(\Sigma, \text{ End } E_\rho)$.

Thus

(12) $$(\eta^{-1}\bar{\partial}\eta)^{\cdot}\langle \alpha \rangle = 0 = (\eta^{-1}\bar{\partial}\eta)^{\cdot}\langle \bar{\alpha} \rangle \, .$$

Analogously

$$(\eta^{-1}\partial\eta)^{\cdot}\langle\alpha\rangle = 0 = (\eta^{-1}\partial\eta)^{\cdot}\langle\bar{\alpha}\rangle\,, \tag{13}$$

thus, $\dot{\eta}$ is constant. Since we require a normalization in z_0, (and since ρ is irreducible) we conclude

$$\dot{\eta} \equiv 0. \tag{14}$$

We now compute mixed second derivatives

$$\begin{aligned} &\partial((\eta^{-1}\bar{\partial}\eta)^{\cdot'}[\mu]\langle\alpha\rangle = 0 && (15)\\ &\partial((\eta^{-1}\bar{\partial}\eta)^{\cdot'}[\mu]\langle\bar{\alpha}\rangle) + (\eta^{-1}\bar{\partial}\eta)^{'}[\mu]\cdot\bar{\alpha}^t(\eta^{-1}\bar{\partial}\eta)^{'}[\mu] = 0 &&\\ \Leftrightarrow\quad &\partial((\eta^{-1}\bar{\partial}\eta)^{\cdot'}[\mu]\langle\bar{\alpha}\rangle + \mu\bar{\alpha}^t - \bar{\alpha}^t\mu = 0\,. && (16)\end{aligned}$$

We look at the transformation behavior of $\dot{\eta}'$:

$$\begin{aligned}\dot{\eta}'(\gamma z) &= \rho'(\gamma)\dot{\eta}(z)\rho(\gamma)^{-1} + \rho(\gamma)\dot{\eta}'(z)\rho(\gamma)^{-1}\\ &= \rho(\gamma)\dot{\eta}'(z)\rho(\gamma)^{-1}, \qquad \text{since } \dot{\eta}\equiv 0 \quad (\text{ cf. } (11)).\end{aligned} \tag{17}$$

Thus, $\dot{\eta}'$ transforms as an infinitesimal automorphism of E_ρ.
(15) then implies as before

$$(\eta^{-1}\bar{\partial}\eta)^{\cdot'}[\mu]\langle\alpha\rangle = 0. \tag{18}$$

Similarly

$$(\eta^{-1}\bar{\partial}\eta)^{\cdot'}[\bar{\mu}]\langle\alpha\rangle = 0 = (\eta^{-1}\bar{\partial}\eta)^{\cdot'}[\bar{\mu}]\langle\bar{\alpha}\rangle\,. \tag{19}$$

Since ρ is irreducible, the only holomorphic sections of End E_ρ are multiples of the unit matrix. We denote by Δ_0 the restriction of the Laplace operator to the orthogonal complement of those.
We rewrite (16) as

$$\partial((\eta^{-1}\bar{\partial}\eta)^{\cdot'}[\mu]\langle\bar{\alpha}\rangle) - \partial\bar{\partial}\Delta_0^{-1}(-\mu\bar{\alpha}^t + \bar{\alpha}^t\mu) = 0. \tag{20}$$

$\mu\bar{\alpha}^t + \bar{\alpha}^t\mu$ is a $(1,1)$-form with values in End E_ρ, und $\Delta_0^{-1}(\mu\bar{\alpha}^t + \bar{\alpha}^t\mu)$ hence is an End E_ρ valued function. Therefore $\bar{\partial}\Delta_0^{-1}(\mu\bar{\alpha}^t + \bar{\alpha}^t\mu)$ is also L^2-orthogonal to $H^{0,1}(\Sigma, \text{End}\, E_\rho)$. Thus (17) implies

$$(\eta^{-1}\bar{\partial}\eta)^{\cdot'}[\mu]\langle\bar{\alpha}\rangle = \bar{\partial}\Delta_0^{-1}(-\mu\bar{\alpha}^t + \bar{\alpha}^t\mu). \tag{21}$$

We now compute derivatives of the metric; we recall

$$g_{\mu\bar\nu}(0) = -\frac{i}{2}\frac{d^2}{d\mu d\bar\nu}\int_\Sigma tr(h^{-1}\bar\partial h \wedge \partial h^{-t}\bar h^{t^{-1}}) \tag{22}$$

$$= -\frac{i}{2}\frac{d^2}{d\mu d\bar\nu}\int_\Sigma tr(g^{-1}\eta^{-1}\bar\partial(\eta g) \wedge \partial(\bar g^t\bar\eta^t)\bar\eta^{t^{-1}}\bar g^{t^{-1}}).$$

Then, using (12), (13) and $(\eta^{-1}\bar\partial\eta)'' = 0$

$$g_{\mu\bar\nu,\alpha}(0) = (\frac{d}{d\alpha}g_{\mu\bar\nu})(0)$$

$$= -\frac{i}{2}\int_\Sigma tr((\eta^{-1}\bar\partial\eta)^{\cdot\prime}[\mu]\langle\alpha\rangle \wedge \bar\nu^t \tag{23}$$

$$+ \mu \wedge ((\partial\bar\eta^t)\bar\eta^{t^{-1}})^{\cdot\prime}[\bar\nu]\langle\alpha\rangle)$$

$$= 0,$$

since $\dot\eta'$ transforms as an infinitesimal endomorphism of E_ρ (cf. (17)), and hence $(\eta^{-1}\bar\partial\eta)^{\cdot\prime}$ is L^2-orthogonal to $H^{0,1}(\Sigma, \mathrm{End}\, E_\rho)$ as before.

For the second derivatives, we get, using (18), (19), (12), (13)

$$g_{\mu\bar\nu,\alpha\bar\beta}(0) = -\frac{i}{2}\int_\Sigma tr((\eta^{-1}\bar\partial\eta)^{\cdot\prime}[\mu]\langle\bar\beta\rangle \wedge (\partial\bar\eta^t\bar\eta^{t^{-1}})^{\cdot\prime}[\bar\nu]\langle\alpha\rangle)$$

$$-\frac{i}{2}\int_\Sigma tr((\eta^{-1}\bar\partial\eta)^{\cdot\cdot\prime}[\mu]\langle\alpha,\bar\beta\rangle \wedge \bar\nu^t) \tag{24}$$

$$-\frac{i}{2}\int_\Sigma tr(\mu \wedge (\partial\bar\eta^t\bar\eta^{t^{-1}})^{\cdot\cdot\prime}[\bar\nu]\langle\alpha,\bar\beta\rangle).$$

We now let

$$k : H \to Gl\,(n, \mathbb{C})$$

be the solution of

$$\partial(k^{-1}\bar\partial k) = 0$$

with the same transformation behavior as η :

$$k(\gamma z) = \rho_{\mu+\alpha}(\gamma z)k(z)\rho(\gamma)^{-1}.$$

Then

$$\eta = kf,$$

where f is an endomorphism of E_ρ.

Since for an endomorphism of E_ρ, $f^{-1}\bar\partial f$ is orthogonal to $H^{0,1}(\Sigma,\ \operatorname{End} E_\rho)$, we can also write

$$g_{\mu\bar\nu}(0) = -\frac{i}{2}\frac{d^2}{d\mu d\bar\nu}\int_\Sigma tr(g^{-1}\eta^{-1}\bar\partial(\eta g)\wedge\partial(\bar g^t\bar k^t)\bar k^{t^{-1}}\bar g^{t^{-1}}). \tag{25}$$

Since $(k^{-1}\bar\partial k)^{\cdot} = 0$, $(k^{-1}\bar\partial k)'[\mu] = \mu$ etc., we obtain

$$g_{\mu\bar\nu,\alpha\bar\beta}(0) = -\frac{i}{2}\int_\Sigma tr((\eta^{-1}\bar\partial\eta)^{\cdot\cdot'}[\mu]\langle\alpha,\bar\beta\rangle\wedge\bar\nu^t). \tag{26}$$

Equating (24) and (26) yields

$$\begin{aligned}&\int_\Sigma tr(\mu\wedge(\partial\bar\eta^t\bar\eta^{t^{-1}})^{\cdot\cdot'}\langle\alpha,\bar\beta\rangle\\ &= -\int_\Sigma tr((\eta^{-1}\bar\partial\eta)^{\cdot'}[\mu]\langle\bar\beta\rangle\wedge(\partial\bar\eta^t\bar\eta^{t^{-1}})^{\cdot'}[\bar\nu]\langle\alpha\rangle),\end{aligned} \tag{27}$$

and hence also

$$\begin{aligned}&\int_\Sigma tr((\eta^{-1}\bar\partial\eta)^{\cdot\cdot'}[\mu]\langle\alpha,\bar\beta\rangle\wedge\bar\nu^t)\\ &= -\int_\Sigma tr((\eta^{-1}\bar\partial\eta)^{\cdot'}[\mu]\langle\bar\beta\rangle\wedge(\partial\bar\eta^t\bar\eta^{t^{-1}})^{\cdot'}[\bar\nu]\langle\alpha\rangle).\end{aligned} \tag{28}$$

Inserting (27), (28) into (24), using (21) and integrating by parts gives

$$\begin{aligned}g_{\mu\bar\nu,\alpha\bar\beta}(0) &= \frac{i}{2}\int_\Sigma tr((\eta^{-1}\bar\partial\eta)^{\cdot'}[\mu]\langle\bar\beta\rangle\wedge(\partial\bar\eta^t\bar\eta^{t'^{-1}})^{\cdot'}[\bar\nu]\langle\alpha\rangle)\\ &= \frac{i}{2}\int_\Sigma tr(\bar\partial\Delta_0^{-1}(-\mu\bar\beta^t+\bar\beta^t\mu)\wedge\partial\Delta_0^{-1}(-\bar\nu^t\alpha+\alpha\bar\nu^t))\\ &= -\frac{i}{2}\int_\Sigma tr((-\mu\bar\beta^t+\bar\beta^t\mu)\wedge\Delta_0^{-1}(-\bar\nu^t\alpha+\alpha\bar\nu^t)).\end{aligned} \tag{29}$$

In a similar vein

$$g_{\mu\nu,\bar\alpha\bar\beta}(0) = 0. \tag{30}$$

We conclude for the curvature tensor at 0

$$\begin{aligned}R_{\mu\bar\nu\alpha\bar\beta} &= \frac{1}{2}g_{\mu\bar\nu,\alpha\bar\beta} + \frac{1}{2}g_{\mu\bar\beta,\alpha\bar\nu}\\ &= -\frac{i}{4}\int_\Sigma tr([\bar\beta^t,\mu]\wedge\Delta_0^{-1}[\alpha,\bar\nu^t]) - \frac{i}{4}\int_\Sigma tr([\bar\nu^t,\mu]\wedge\Delta_0^{-1}[\alpha,\bar\beta^t]).\end{aligned} \tag{31}$$

§ 3 The effect of variations of the complex structure of the underlying Riemann surface

We now want to study the effect of variations of Σ. We consider the fiber space N_p over Teichmüller space T_p of Riemann surfaces of genus $p \geq 2$, where the fiber over $\sigma \in T_p$ is the moduli space of vector bundles of rank n and degree k over the Riemann surface Σ with conformal structure determined by σ. Obviously, this also induces a fiber space over moduli space M_p.

The vertical bundle, i.e. the one consisting of the tangent space of the fibers, then has metric $\langle \mu, \nu \rangle$, where $\mu, \nu \in H^{0,1}(\Sigma, \operatorname{End} E_\rho)$ as above. We now study this metric in its dependence on Σ.

The tangent space of T_p at σ is given by harmonic Beltrami differentials on the corresponding surface Σ (cf.[7] p. 195). We denote the space of harmonic Beltrami differentials on Σ by $\mathcal{H}(\Sigma)$. The definition of a harmonic Beltrami differential requires the choice of a metric on Σ. We shall choose the hyperbolic one; the choice of metric, however, will only play an auxiliary rôle.

We return to our functional

$$E(h) = -\frac{i}{2} \int_\Sigma tr(h^{-1}\bar{\partial} h \wedge \overline{(h^{-1}\bar{\partial} h)}^t.$$

For given $\mu \in H^{0,1}(\Sigma, \operatorname{End} E_\rho)$, we let h be the solution of

$$\begin{aligned} h : H &\to Gl(n, \mathbb{C}) \\ h^{-1}\bar{\partial} h &= \mu \\ h(\gamma z) &= \rho_\mu(\gamma) h(z) \rho(\gamma)^{-1} \quad \text{for } \gamma \in \Gamma, z \in H, \end{aligned}$$

as before. In our previous notation

$$h'_{\bar{z}}[\mu] d\bar{z} = \mu, \quad h'_z[\bar{\mu}] dz = -\bar{\mu}^t. \tag{32}$$

We therefore write

$$g_{\mu\bar{\nu}} = \langle \mu, \nu \rangle = \frac{i}{2} \int_\Sigma tr(h'_{\bar{z}}[\mu] \cdot h'_z[\bar{\nu}]) d\bar{z} \wedge dz. \tag{33}$$

Harmonic Beltrami differentials will be denoted by δ, θ.
Given δ, there is a smooth family $\zeta_t : H \to H$ of diffeomorphisms solving

$$\zeta_{t,\bar{z}} = t\delta \zeta_{t,z}, \tag{34}$$

normalized with

$$\zeta_0(z) = z.$$

Then

$$\omega_{\bar{z}} := \frac{\partial}{\partial t} \zeta_{t,\bar{z}|_{t=0}} = \delta. \tag{35}$$

If ζ_t solves (34), so does $\zeta_t \circ \gamma$ for any $\gamma \in \Gamma$, because δ transforms via

$$(\delta \circ \gamma)\bar{\gamma}_{\bar{z}} = \delta \gamma_z. \tag{36}$$

Consequently, there exists a group Γ^t operating conformally on H, with unique fix point z_0, with the property that for every $\gamma \in \Gamma$, there exists $\gamma^t \in \Gamma^t$ with

$$\zeta_t \circ \gamma = \gamma^t \circ \zeta_t. \tag{37}$$

$\Sigma^t = H/\Gamma^t$ then defines the family of surfaces induced by the family $t\delta$.

Now we want to construct a holomorphic section of the fiber bundle N_p over T_p (c.f. [6]). We fix a neighborhood U of the origin in $\mathcal{H}(\Sigma)$, the tangent space of T_p at Σ, such that for $\delta \in U$

$$|\delta|_{L^\infty} < 1.$$

Let ζ_t be the smooth family of diffeomorphisms of H defined by the equation (31). Then, by a theorem of Bers [3], the map

$$\begin{aligned} \varphi : U \times N(n,k,\Sigma) &\longrightarrow N_p \\ (\delta, E_\rho) &\longmapsto (\delta, E_\rho^t) \end{aligned}$$

is holomorphic in δ (but not in E_ρ) where E_ρ^t is defined as follows:
We consider

$$\begin{aligned} H \times \mathbf{C}^n &\longrightarrow H \times \mathbf{C}^n \\ (z,x) &\longmapsto (\zeta_t(z), x), \end{aligned}$$

and take the quotient by Γ on the left and by Γ^t on the right and get

$$\begin{array}{ccc} E_\rho & \longrightarrow & E_\rho^t \\ \downarrow & & \downarrow \\ \Sigma & \longrightarrow & \Sigma^t, \end{array}$$

since

$$\begin{aligned} (z,x) &\sim (\gamma z, \rho(\gamma)x) \\ (\zeta_t(z), x) &\sim (\gamma^t \zeta_t(z), \rho(\gamma^t)x) = (\zeta_t\gamma(z), \rho(\gamma^t)x). \end{aligned}$$

(Note: ρ is a representation of the abstract group Γ in $U(n)$, and as abstract group Γ^t is isomorphic to Γ, and therefore ρ defines also a representation of Γ^t.)

If we fix E_ρ, the map φ gives a local holomorphic section of N_p. For any section F of the vertical bundle depending on δ, we denote its corresponding derivative by

$$F'\{\delta\}.$$

It means that we first pull F back to E_ρ over Σ using ζ_t, and then differentiate it w.r.t. t.

Since for $\mu = 0$, the solution h of our equation is the unit matrix E, we have

$$h'\{\delta\} = 0 \quad \text{at } \mu = 0. \tag{38}$$

Consequently, we have for the transformation behaviour of

$$h'' = h''[\mu]\{\delta\}$$

$$h''(\gamma z) = \rho(\gamma)h''(z)\rho(\gamma)^{-1}. \tag{39}$$

Thus, h'' transforms as an infinitesimal automorphism of E_ρ, and we may integrate by parts in the sequel.
This implies in particular

$$\int_\Sigma tr(h''_{\bar z}[\mu]\{\delta\} \cdot h'_z[\bar\nu])d\bar z \wedge dz = 0, \tag{40}$$

because $h'_z[\bar\nu] = -\bar\nu^t$ and $\bar\nu_{\bar z} = 0$, since $\nu \in H^{0,1}(\Sigma, \operatorname{End} E_\rho)$, and likewise

$$\int_\Sigma tr(h'_{\bar z}[\mu] \cdot h''_z[\bar\nu]\{\delta\})d\bar z \wedge dz = 0. \tag{41}$$

Consequently, the derivative of $\langle \mu, \nu \rangle$ in the direction δ is given by

$$\frac{i}{2}\frac{d}{dt}\int_{\Sigma^t} tr(h'_{\bar\zeta}[\mu] \cdot h'_\zeta[\bar\nu])d\bar\zeta \wedge d\zeta_{|t=0} \tag{42}$$

with $\zeta = \zeta_t$.

From the chain rule, we compute

$$\begin{aligned}
&-\int_{\Sigma^t} tr(h'_{\bar\zeta}[\mu] \cdot h'_\zeta[\bar\nu])d\bar\zeta \wedge d\zeta \\
&= -\int_{\Sigma^t} tr((h'_z[\mu]z_{\bar\zeta} + h'_{\bar z}[\mu]\bar z_{\bar\zeta}) \cdot (h'_z[\bar\nu]z_\zeta + h'_{\bar z}[\bar\nu]\bar z_\zeta))d\bar\zeta \wedge d\zeta \\
&= -\int_\Sigma tr(h'_{\bar z}[\mu] \cdot h'_z[\bar\nu]\zeta_z\bar\zeta_{\bar z} + h'_z[\mu] \cdot h'_{\bar z}[\bar\nu]\zeta_{\bar z}\bar\zeta_z \\
&\quad - h'_{\bar z}[\mu] \cdot h'_{\bar z}[\bar\nu]\zeta_z\bar\zeta_z - h'_z[\mu] \cdot h'_z[\bar\nu]\zeta_{\bar z}\bar\zeta_{\bar z})\frac{d\bar z \wedge dz}{\zeta_z\bar\zeta_{\bar z} - \bar\zeta_z\zeta_{\bar z}}.
\end{aligned} \tag{43}$$

Using (35), (40), (41), we obtain

$$\begin{aligned}
&-\frac{d}{dt}\int_{\Sigma^t} tr(h'_{\bar\zeta}[\mu] \cdot h'_\zeta[\bar\nu])d\bar\zeta \wedge d\zeta_{|t=0} \\
&= \int_\Sigma tr(h'_{\bar z}[\mu] \cdot h'_{\bar z}[\bar\nu])\bar\delta d\bar z \wedge d\bar z + \int_\Sigma tr(h'_z[\mu] \cdot h'_z[\bar\nu])\delta\, dz \wedge dz \\
&= 0, \quad \text{since } h'_{\bar z}[\bar\nu] = h'_z[\mu] = 0,
\end{aligned}$$

(note that δ transforms like $\frac{\partial}{\partial z} \otimes d\bar{z}$, so that we integrate (1,1)-forms). Thus

$$g_{\mu\bar{\nu},\delta} = 0 = g_{\mu\bar{\nu},\bar{\delta}}. \tag{44}$$

Likewise, we can define

$$g_{\mu\nu} := \frac{i}{2} \int_{\Sigma} tr(h'_{\bar{z}}[\mu] \cdot h'_{z}[\nu]) d\bar{z} \wedge dz,$$

and we obtain

$$g_{\mu\nu,\delta} = 0\,, \tag{45}$$

$$\begin{aligned} g_{\mu\nu,\bar{\delta}} &= -\frac{i}{2} \int_{\Sigma} tr(h'_{\bar{z}}[\mu] \cdot h'_{\bar{z}}[\nu]) \bar{\delta} d\bar{z} \wedge d\bar{z} \\ &= -\frac{i}{2} \int_{\Sigma} tr(\mu \wedge \nu) \bar{\delta}\,. \end{aligned} \tag{46}$$

We now turn to second derivatives. As a preliminary, we study variations of the solution of our differential equation.
We write the solution $w : H \to Gl(n, \mathbb{C})$ of

$$\begin{aligned} &\partial(w^{-1} \bar{\partial} w) = 0 \\ &w(\gamma^t z) = \rho_\mu(\gamma^t) w(z) \rho(\gamma^t)^{-1} \quad \text{for } \gamma^t \in \Gamma^t, z \in H \end{aligned} \tag{47}$$

as

$$w(\zeta) = h(z(\zeta)),$$

where h is as above, and $\zeta = \zeta_t$ as before.
(47) then is transformed into

$$\begin{aligned} 0 =& h_z \frac{1}{\zeta_z \bar{\zeta}_{\bar{z}} - \zeta_{\bar{z}} \bar{\zeta}_z} (\zeta_{zz} \bar{\zeta}_{\bar{z}} \bar{\zeta}_{\bar{z}} \zeta_{\bar{z}} + \zeta_{\bar{z}\bar{z}} \bar{\zeta}_{\bar{z}} \bar{\zeta}_z \zeta_z - \bar{\zeta}_{zz} \zeta_{\bar{z}} \zeta_{\bar{z}} \bar{\zeta}_{\bar{\zeta}} \\ & - \bar{\zeta}_{\bar{z}\bar{z}} \zeta_{\bar{z}} \bar{\zeta}_z \zeta_z + (\bar{\zeta}_{z\bar{z}} \zeta_{\bar{z}} - \zeta_{\bar{z}z} \bar{\zeta}_{\bar{\zeta}})(\zeta_z \bar{\zeta}_{\bar{z}} + \zeta_{\bar{z}} \bar{\zeta}_z)) \\ & + h_{\bar{z}} \frac{1}{\zeta_z \bar{\zeta}_{\bar{z}} - \zeta_{\bar{z}} \bar{\zeta}_z} (\bar{\zeta}_{\bar{z}\bar{z}} \zeta_z \zeta_z \bar{\zeta}_z + \bar{\zeta}_{zz} \zeta_z \zeta_{\bar{z}} \bar{\zeta}_{\bar{z}} - \zeta_{\bar{z}\bar{z}} \bar{\zeta}_z \bar{\zeta}_z \zeta_z \\ & - \zeta_{zz} \bar{\zeta}_z \zeta_{\bar{z}} \bar{\zeta}_{\bar{z}} + (\zeta_{\bar{z}z} \bar{\zeta}_z - \bar{\zeta}_{z\bar{z}} \zeta_z)(\zeta_z \bar{\zeta}_{\bar{z}} + \zeta_{\bar{z}} \bar{\zeta}_z)) \\ & - (h_{zz} - h_z h^{-1} h_z) \zeta_{\bar{z}} \bar{\zeta}_{\bar{z}} - (h_{\bar{z}\bar{z}} - h_{\bar{z}} h^{-1} h_{\bar{z}}) \zeta_z \bar{\zeta}_z \\ & + (h_{z\bar{z}} - h_z h^{-1} h_{\bar{z}})(\zeta_z \bar{\zeta}_{\bar{z}} + \zeta_{\bar{z}} \bar{\zeta}_z). \end{aligned} \tag{48}$$

Differentiating (48) at $t = 0$ and noting that at $t = 0$, $\zeta(z) = z$, we obtain

$$\begin{aligned} 0 &= -h'_z[\mu] \delta_z - h'_{zz}[\mu] \delta + h''_{z\bar{z}}[\mu]\{\delta\} \\ &= h''_{z\bar{z}}[\mu]\{\delta\}, \quad \text{since} \quad h'_z[\mu] = 0. \end{aligned} \tag{49}$$

Since, by (39), h' represents an infinitesimal automorphism of E_ρ, we conclude

$$h'_z{}'[\mu]\{\delta\} = 0 = h'_{\bar z}{}'[\mu]\{\delta\}. \tag{50}$$

Similarly

$$0 = -h'_z[\bar\nu]\delta_z - h'_{zz}[\bar\nu]\delta + h'_{z\bar z}{}'[\bar\nu]\{\delta\}. \tag{51}$$

Thus,

$$\frac{\partial}{\partial z}(h'_{\bar z}{}'[\bar\nu]\{\delta\} - h'_z[\bar\nu]\delta) = 0. \tag{52}$$

Also

$$\begin{aligned} \frac{\partial}{\partial \bar z}(h'_z{}'[\bar\nu]\{\delta\}) &= \frac{\partial}{\partial z}(h'_z[\bar\nu]\delta) \\ &= -\frac{\partial}{\partial z}\bar\nu^t\delta \\ &= -\frac{\partial}{\partial z}\frac{\partial^2}{\partial z\partial\bar z}\Delta_0^{-1}(\bar\nu^t\delta). \end{aligned} \tag{53}$$

Again, $\frac{\partial}{\partial z}\Delta_0^{-1}(\bar\nu^t\delta)$ transforms as an $\mathrm{End}E_\rho$-valued function. The same is true for $h'_z{}'[\bar\nu]\{\delta\}$, cf.(39). Thus, they are both L^2-orthogonal to $H^{0,1}(\Sigma, \mathrm{End}E_\rho)$, and we conclude

$$h'_z{}'[\bar\nu]\{\delta\} = -\frac{\partial^2}{\partial z^2}\Delta_0^{-1}(\bar\nu^t\delta). \tag{54}$$

We now compute from (43), using (50)

$$\begin{aligned} g_{\mu\bar\nu,\delta\bar\theta} =& \frac{i}{2}\int_\Sigma tr(h'_{\bar z}{}''[\mu]\{\delta,\bar\theta\}\cdot h'_z[\bar\nu])d\bar z\wedge dz \\ &+\frac{i}{2}\int_\Sigma tr(h'_{\bar z}[\mu]\cdot h'_z{}''[\bar\nu]\{\delta,\bar\theta\})d\bar z\wedge dz \\ &+\frac{i}{2}\int_\Sigma tr(h'_{\bar z}{}'[\mu]\{\bar\theta\}\cdot h'_z{}'[\bar\nu]\{\delta\})d\bar z\wedge dz \\ &+\frac{i}{2}\int_\Sigma tr(h'_{\bar z}{}'[\bar\nu]\{\delta\}\cdot h'_z{}'[\mu]\{\bar\theta\})d\bar z\wedge dz \\ &-\frac{i}{2}\int_\Sigma tr(h'_{\bar z}[\mu]\cdot h'_{\bar z}{}'[\bar\nu]\{\delta\})\bar\theta d\bar z\wedge dz \\ &-\frac{i}{2}\int_\Sigma tr(h'_z{}'[\mu]\{\bar\theta\}\cdot h'_z[\bar\nu])\delta d\bar z\wedge dz \\ &+\frac{i}{2}\int_\Sigma tr(h'_{\bar z}[\mu]\cdot h'_z[\bar\nu])\delta\bar\theta d\bar z\wedge dz. \end{aligned} \tag{55}$$

We recall that h has to be considered as the solution of our equation on the varying surface Σ^t. We let l be the solution on the fixed surface Σ. Then h and l differ only by a diffeomorphism of Σ, and since infinitesimal diffeomorphisms are L^2-orthogonal to harmonic forms, we can also write

$$g_{\mu\bar{\nu}} = \frac{i}{2}\int_\Sigma tr(h'_{\bar{z}}[\mu]\cdot l'_z[\bar{\nu}])d\bar{z}\wedge dz. \tag{56}$$

This yields

$$\begin{aligned} g_{\mu\bar{\nu},\delta\bar{\theta}} =& \frac{i}{2}\int_\Sigma tr(h_{\bar{z}}^{'''}[\mu]\{\delta,\bar{\theta}\}\cdot l'_z[\bar{\nu}])d\bar{z}\wedge dz \\ &-\frac{i}{2}\int_\Sigma tr(h_{\bar{z}}^{''}[\mu]\{\bar{\theta}\}\cdot l'_z[\bar{\nu}])\delta d\bar{z}\wedge dz \\ &+\frac{i}{2}\int_\Sigma tr(h'_{\bar{z}}[\mu]\cdot l'_z[\bar{\nu}])\delta\bar{\theta} d\bar{z}\wedge dz. \end{aligned} \tag{57}$$

Of course $h'_{\bar{z}}[\mu]=\mu$, $l'_z[\bar{\nu}]=h'_z[\bar{\nu}]=-\bar{\nu}^t$. Equating (55) and (57) then gives

$$\begin{aligned} 0 =& -\int_\Sigma tr(h'_{\bar{z}}[\mu]\cdot h_z^{'''}[\bar{\nu}]\{\delta,\bar{\theta}\})d\bar{z}\wedge dz \\ &-\int_\Sigma tr(h_{\bar{z}}^{''}[\mu]\{\bar{\theta}\}\cdot h_z^{''}[\bar{\nu}]\{\delta\})d\bar{z}\wedge dz \\ &-\int_\Sigma tr(h_{\bar{z}}^{''}[\nu]\{\delta\}\cdot h_z^{''}[\bar{\mu}]\{\bar{\theta}\})d\bar{z}\wedge dz \\ &+\int_\Sigma tr(h'_{\bar{z}}[\mu]\cdot h_{\bar{z}}^{''}[\bar{\nu}]\{\delta\})\bar{\theta} d\bar{z}\wedge dz. \end{aligned} \tag{58}$$

Conjugating (58) and inserting the resulting expression into (57) gives

$$\begin{aligned} g_{\mu\bar{\nu},\delta\bar{\theta}} =& \frac{i}{2}\int_\Sigma tr(h_{\bar{z}}^{''}[\mu]\{\bar{\theta}\}\cdot h_z^{''}[\bar{\nu}]\{\delta\})d\bar{z}\wedge dz \\ &-\frac{i}{2}\int_\Sigma tr(h_{\bar{z}}^{''}[\bar{\nu}]\{\delta\}\cdot h_z^{''}[\mu]\{\bar{\theta}\})d\bar{z}\wedge dz \\ &+\frac{i}{2}\int_\Sigma tr(h'_{\bar{z}}[\mu]\cdot h'_z[\bar{\nu}])\delta\bar{\theta} d\bar{z}\wedge dz. \end{aligned}$$

Using (52), (54) results in

$$\begin{aligned} g_{\mu\bar{\nu},\delta\bar{\theta}} =& -\frac{i}{2}\int_\Sigma tr(\frac{\partial^2}{\partial\bar{z}^2}\Delta_0^{-1}(\mu\bar{\theta})\wedge\frac{\partial^2}{\partial z^2}\Delta_0^{-1}(\bar{\nu}^t\delta)) \\ &+\frac{i}{2}\int_\Sigma tr(\bar{\nu}^t\delta\wedge\mu\bar{\theta})-\frac{i}{2}\int_\Sigma tr(\mu\wedge\bar{\nu}^t)\delta\bar{\theta} \\ =& -\frac{i}{2}\int_\Sigma tr((\mu\bar{\theta})\wedge(\bar{\nu}^t\delta)). \end{aligned} \tag{59}$$

Similarly,

$$
\begin{aligned}
g_{\mu\nu,\bar\delta\bar\theta} =& -\frac{i}{2}\int_\Sigma tr(h'_{\bar z}[\mu]\{\bar\delta\}\cdot l'_{\bar z}[\nu])\bar\theta d\bar z\wedge dz \\
& -\frac{i}{2}\int_\Sigma tr(h''_{\bar z}[\mu]\{\bar\theta\}\cdot l'_{\bar z}[\nu])\bar\delta d\bar z\wedge dz \\
=& \frac{i}{2}\int_\Sigma tr(\frac{\partial^2}{\partial\bar z^2}(\Delta_0^{-1}(\mu\bar\delta))\wedge\nu\bar\theta) \\
& +\frac{i}{2}\int_\Sigma tr(\frac{\partial^2}{\partial\bar z^2}(\Delta_0^{-1}(\mu\bar\theta))\wedge\nu\bar\delta).
\end{aligned}
\tag{60}
$$

Finally, we want to compute mixed derivative of $g_{\mu\bar\nu}$, i.e. differentiating once in the fiber and in the base direction.
From (40) again

$$
\begin{aligned}
g_{\mu\bar\nu,\alpha\bar\delta} =& \frac{i}{2}\int_\Sigma tr(h'''_{\bar z}[\mu,\alpha]\{\bar\delta\}\cdot h'_{z}[\bar\nu])d\bar z\wedge dz \\
& +\frac{i}{2}\int_\Sigma tr(h'_{\bar z}[\mu]\cdot h'''_{z}[\bar\nu,\alpha]\{\bar\delta\})d\bar z\wedge dz \\
& +\frac{i}{2}\int_\Sigma tr(h''_{\bar z}[\mu]\langle\alpha\rangle\{\bar\delta\}\cdot h'_{z}[\bar\nu])d\bar z\wedge dz \\
& +\frac{i}{2}\int_\Sigma tr(h'_{\bar z}[\mu]\cdot h''_{z}[\bar\nu]\langle\alpha\rangle\{\bar\delta\})d\bar z\wedge dz \\
& +\frac{i}{2}\int_\Sigma tr(h''_{\bar z}[\mu]\{\bar\delta\}\cdot h'_{z}[\bar\nu]\langle\alpha\rangle)d\bar z\wedge dz \\
& +\frac{i}{2}\int_\Sigma tr(h'_{\bar z}[\mu]\langle\alpha\rangle\cdot h''_{z}[\bar\nu]\{\bar\delta\})d\bar z\wedge dz \\
& -\frac{i}{2}\int_\Sigma tr(h'_{\bar z}[\mu]\langle\alpha\rangle\cdot h'_{\bar z}[\bar\nu])\bar\delta d\bar z\wedge dz \\
& -\frac{i}{2}\int_\Sigma tr(h'_{\bar z}[\mu]\cdot h'_{\bar z}[\bar\nu]\langle\alpha\rangle)\bar\delta d\bar z\wedge dz,
\end{aligned}
\tag{61}
$$

and as before also

$$
\begin{aligned}
g_{\mu\bar\nu,\alpha\bar\delta} =& \frac{i}{2}\int_\Sigma tr(h'''_{\bar z}[\mu,\alpha]\{\bar\delta\}\cdot l'_{z}[\bar\nu])d\bar z\wedge dz \\
& +\frac{i}{2}\int_\Sigma tr(h''_{\bar z}[\mu]\langle\alpha\rangle\{\bar\delta\}\cdot l'_{z}[\bar\nu])d\bar z\wedge dz \\
& +\frac{i}{2}\int_\Sigma tr(h''_{\bar z}[\mu]\{\bar\delta\}\cdot l'_{z}[\bar\nu]\langle\alpha\rangle)d\bar z\wedge dz \\
& -\frac{i}{2}\int_\Sigma tr(h'_{\bar z}[\mu]\langle\alpha\rangle\cdot l'_{\bar z}[\bar\nu])\bar\delta d\bar z\wedge dz.
\end{aligned}
\tag{62}
$$

From (18), (19) and $h'_{\bar z}[\bar\nu]=0$, we conclude that the last three terms in (61) vanish, and so does the term in (62).

Equating (61) and (62) then yields

$$
0 = -\int_{\Sigma} tr(h'_{\bar{z}}[\mu] \cdot h''^{\prime}_{z}[\bar{\nu},\alpha]\{\bar{\delta}\})d\bar{z}\wedge dz
$$

(63)

$$
-\int_{\Sigma} tr(h'_{\bar{z}}[\mu] \cdot h'^{\cdot\prime}_{z}[\bar{\nu}]\langle\alpha\rangle\{\bar{\delta}\})d\bar{z}\wedge dz
$$

$$
-\int_{\Sigma} tr(h'^{\prime}_{\bar{z}}[\mu]\{\bar{\delta}\} \cdot h'^{\cdot}_{z}[\bar{\nu}]\langle\alpha\rangle)d\bar{z}\wedge dz.
$$

Using this in (62) gives

$$
g_{\mu\bar{\nu},\alpha\bar{\delta}} = -\frac{i}{2}\int_{\Sigma} tr(h'^{\prime}_{\bar{z}}[\mu]\{\bar{\delta}\} \cdot h'^{\cdot}_{z}[\bar{\nu}]\langle\alpha\rangle)d\bar{z}\wedge dz
+\frac{i}{2}\int_{\Sigma} tr(h'^{\prime}_{\bar{z}}[\mu]\{\bar{\delta}\} \cdot l'^{\cdot}_{z}[\bar{\nu}]\langle\alpha\rangle)d\bar{z}\wedge dz,
$$

since $h^{\cdot\prime} = l^{\cdot\prime}$. Hence

(64)

$$
g_{\mu\bar{\nu},\alpha\bar{\delta}} = 0.
$$

References

[1] Axelrod, S., della Pietra and Witten, E., Geometric quantization of Chern Simons gauge theory, preprint

[2] Beilinson, A. and Kazhdan, D., Flat projective connections, preprint

[3] Bers, L., Fibre spaces over Teichmüller spaces, Acta Math. 130 (1973) 89-126

[4] Hitchin, N.J., Flat connections and geometric quantization, preprint.

[5] Jost, J., Nonlinear Methods in Riemannian and Kählerian Geometry, DMV Seminar Band 10, Birkhäuser-Verlag,Basel-Boston (1988)

[6] Jost, J., Harmonic maps and curvature computations in Teichmüller theory, Ann. Acad. Sci. Fenn., to appear

[7] Jost, J., Two Dimensional Geometric Variational Problems, Wiley-Interscience, Chichester 1991

[8] Kobayashi, S., Differential Geometry of Complex Vector Bundles, Publications of the mathematical society of Japan, vol.15, Iwanami Shoten Publishers and Princeton University Press (1987)

[9] Narasimhan, M.S. and Seshadri, C.S., Holomorphic vector bundles on a compact Riemann surface, Math. Ann. 155 (1964) 69-80

[10] Narasimhan, M.S. and Seshadri, C.S., Stable and unitary vector bundles on a compact Riemann surface, Ann of Math. 82 (1965) 540-569

[11] Rauch, H.E., A transcendental view of the space of algebraic Riemann surfaces, Bull. Amer. Math. Soc.71 (1965) 1-39

[12] Takhtajan, L.A., Uniformization, local index theorem, and geometry of the moduli spaces of Riemann surfaces and vector bundles, Proc. of Symposia in Pure Math. vol. 49 (1989) Part I, 581-596

[13] Zograf, P.G. and Takhtadzhyan, L.A., Narasimhan-Seshadri connection and Kähler structure of the space of moduli of holomorphic vector bundles over Riemann surfaces, Fct. Anal. Appl. 20 (1986) 240-241

[14] Zograf, P.G. and Takhtadzhyan, L.A., On geometry of the moduli spaces of vector bundles over a Riemann surfaces , Izv. Akad. Nauk USSR, ser. Mat. 53 (1989) (in Russian)

Mathematisches Institut
Ruhr-Universität Bochum
Universitätsstraße 150

D-4630 Bochum

A CANONICAL CONNECTION FOR LOCALLY HOMOGENEOUS RIEMANNIAN MANIFOLDS

O. KOWALSKI AND F. TRICERRI

1. Introduction

Let (M,g) be a Riemannian manifold of dimension n. We say that (M,g) is *locally homogeneous* if the pseudogroup of the local isometries is transitive on M, i.e. if for each pair of points p and q of M there exists a neighbourhood U of p, a neighbourhood U' of q and an isometry $f : U \to U'$ sending p in q.

Following I.M. Singer [SI], we define k_M as the first integer for which we have

$$(1.1) \qquad \mathbf{g}(q;k_M) = \mathbf{g}(q;k_M+1),$$

where $\mathbf{g}(q;s)$ is the Lie algebra of the skew-symmetric endomorphisms of T_qM which annihilate the Riemann curvature tensor R and its covariant derivatives D^rR, evaluated at q, up to the order s. That is

$$(1.2) \qquad \mathbf{g}(q;s) = \{ A \in \mathbf{so}(T_qM) : A \cdot D^rR_{|q} = 0, 0 \le r \le s \}.$$

Here, we put $D^0R = R$ and we denote by $\mathbf{so}(T_qM)$ the Lie algebra of the skew-symmetric endomorphisms of T_qM. Note that $k_M+1 < (3/2)n$ (cfr. [GR], p.165).

Actually, the Lie algebras $\mathbf{g}(q;s)$ are all isomorphic for different points q of M. Nevertheless, it is better for our purposes to avoid any identification and to keep distinct these algebras.

We put

$$(1.3) \qquad \mathbf{h}(q) = \mathbf{g}(q;k_M).$$

All these algebras fit together in a *vector bundle* E over M (cfr. Lemma 2.1). E is a subbundle of $\mathbf{so}(M)$, where $\mathbf{so}(M)$ is the vector fibre bundle over M whose fibres are the vector spaces $\mathbf{so}(T_qM)$. The Riemann metric g of M induces a fibre metric on $\mathbf{so}(M)$ (i.e. an inner product on the fibres which depends smoothly on q) given by

$$\bar{g}(A,B) = \sum_{i=1}^{n} g_q(A(e_i),B(e_i))$$

for each $A,B \in \mathbf{so}(T_qM)$, where $(e_1,\dots,e_n)$ is an orthonormal basis of T_qM. Of course, the definition does not depend on the choice of the orthonormal basis. Moreover, the restriction of $\overline{g}$ to the fibres $\mathbf{so}(T_qM)$ of $\mathbf{so}(M)$ coincides, up to a constant factor, with the *Killing form* of $\mathrm{so}(T_qM)$.

Since $\mathbf{so}(M)$ carries the metric $\overline{g}$, it splits into the direct sum of E and of its orthogonal complement $E^{\perp}$. By keeping in mind this splitting, we can state now the main theorem of this paper :

Theorem 1.1. Let (M,g) be a locally homogeneous Riemannian manifold, then there exists a unique metric connection ∇ on (M,g) such that :

a) the torsion and the curvature of ∇ are parallel with respect to ∇.

b) Let S be the difference tensor field between ∇ and the Levi Civita connection D, then, for each vector field X on M, the operator S_X defined by

$$S_X : Y \to S_X Y,$$

Y any vector field on M, is a section of $E^{\perp}$.

The connection ∇ is a purely Riemannian invariant (cfr. Remark 2.1). It depends only on the Riemannian structure of M. In fact we have :

Theorem 1.2. The tensor field S at a point q is uniquely determined by the following two conditions :

$$i_X D^{s+1} R_{|q} = -\ (S_q)_X \cdot D^s R_{|q} \tag{1.5}$$

for all $s \leq k_M$ (i_X denotes the interior product by X) and

$$(S_q)_X \in h(q)^{\perp}, \tag{1.6}$$

where $h(q)^{\perp}$ is the orthogonal complement of $h(q)$ in $\mathrm{so}(T_qM)$ with respect to the metric $\overline{g}$ (it is the fibre of $E^{\perp}$ over q).

If (M,g) is *locally symmetric*, then the connection ∇ coincides with the Levi Civita connection D. In fact, in this case we have $k_M = 0$ since $DR = 0$. Therefore, from Theorem 1.2 we get that S_q is also an element of $\mathbf{h}(q)$ for all $q \in M$. So, it is zero and $\nabla = D$.

The connection ∇ has implicitly been constructed in [NT] in order to give an alternative proof of a theorem of I.M. Singer stating that an *infinitesimally homogeneous* Riemannian manifold is locally homogeneous.

For some other applications of this connection we refer the reader to [LA] and [LT].

2. The proof of the main theorem

In order to prove Theorem 1.1 we need two lemmas.

Lemma 2.1. The set $E = \bigcup_{q\in M} \mathbf{h}(q)$ is a vector subbundle of $\mathbf{so}(M)$.

Proof. We have to show that each point q of M has a neighbourhood U where $n(n-1)/2$ linearly independent smooth sections of $\mathbf{so}(M)$ are defined in such a way that the first $r = \dim\mathbf{h}(q)$ sections give a basis of $\mathbf{h}(x)$, for each $x \in U$.

In order to do this, we consider a linear metric connection ∇ such that $\nabla_X D^s R = 0$ for $0 \le s \le k_M+1$. It exists because (M,g) is locally homogeneous (cfr. [NT], [SI]). Let U be a normal neighbourhood of q (with respect to ∇) and $(e_1,\dots,e_n)$ an orthonormal basis of T_qM. Let $E_1,\dots,E_n$ be the vector fields obtained by parallel transport (with respect to ∇) along the ∇-geodesics through q. We get in this way a local orthonormal frame $(E_1,\dots,E_n)$ defined on U. Then, we choose a basis $H_1(q),\dots,H_r(q)$ of $\mathbf{h}(q)$. We complete it by $H_{r+1}(q),\dots,H_{n(n-1)/2}(q)$ in order to get a basis of $\mathbf{so}(T_qM)$. We transport this basis by parallelism (with respect to ∇) along the ∇-geodesics, i.e. if

$$H_\alpha(q)(e_i) = \sum_{j=1}^{n} h_{\alpha ij}\, e_j,$$

we put

$$H_\alpha(x)(E_i(x)) = \sum_{j=1}^{n} h_{\alpha ij}\, E_j(x).$$

Since $\nabla_X D^s R = 0$, the components of D^sR with respect to the local frame $(E_1,\dots,E_n)$ are constant. Then, it is easy to prove that the operators $H_\alpha(x)$ belong to $\mathbf{h}(x)$ if $1 \le \alpha \le r = \mathit{dim}\mathbf{h}(q)$. It follows that they are the elements of a basis of $\mathbf{h}(x)$, because $\mathit{dim}\mathbf{h}(x) = \mathit{dim}\mathbf{h}(q)$. So, $H_1,\dots,H_r,H_{r+1},\dots,H_{n(n-1)/2}$ are the desired local smooth sections of $\mathbf{so}(M)$.

Q.E.D.

Now, let $\mathbf{so}(M) = E \oplus E^\perp$ be the orthogonal splitting of $\mathbf{so}(M)$ with respect to the metric $\bar{g}$ introduced in section 1. Then we have :

Lemma 2.2. Let ∇ be any linear metric connection on M such that $\nabla_X D^s R = 0$ for $0 \leq s \leq k_M+1$, then the covariant derivative $\nabla_X \sigma$ of a smooth section σ of E (resp. $E^\perp$), with respect to any vector field X, is still a section of E (resp. $E^\perp$).

Proof. Let U and $(E_1,\dots,E_n)$ be as in the proof of Lemma 2.1. Let Φ be the **so**(M)-valued local 1-form defined by

$$\Phi(X)(E_i) = \nabla_X E_i.$$

Actually, Φ is E-valued. In fact, from $\nabla_X D^s R = 0$ we get

$$\Phi(X)\cdot D^s R = 0$$

for $0 \leq s \leq k_M+1$, since the components of $D^s R$ with respect to the frame $(E_1,\dots,E_n)$ are constant. Therefore, $\Phi_{|x}$ belongs to the fibre $\mathbf{h}(x)$ of E, for all $x \in U$.

Since there is a local basis of E (resp. of $E^\perp$) whose components with respect to $(E_1,\dots,E_n)$ are constants (use the H_α above), it is enough to prove the lemma for sections with the same property. If σ is such a section, we have

$$(\nabla_X \sigma)(E_i) = \Phi(X)(\sigma(E_i)) - \sigma(\Phi(X)(E_i)).$$

So,

$$\nabla_X \sigma = \Phi(X)\circ\sigma - \sigma\circ\Phi(X) = \mathrm{ad}(\Phi(X))(\sigma).$$

We have already remarked that the restriction of the metric $\overline{g}$ to the fibres of **so**(M) coincides, up to a constant factor, with the Killing form of the Lie algebra $\mathbf{so}(T_qM)$. Since $\mathbf{h}(x)$ is a Lie subalgebra of $\mathbf{so}(T_xM)$, we get in particular that both $\mathbf{h}(x)$ and $\mathbf{h}(x)^\perp$ are $\mathrm{ad}(\mathbf{h}(x))$-invariant for all $x \in U$. This proves that $\nabla_X \sigma$ is still a section of E (resp. $E^\perp$).

Q.E.D.

Now, we are ready to prove Theorem 1.1 and Theorem 1.2.

Proof of Theorem 1.1. First, let us prove the *uniqueness* of the connection. Let ∇ and ∇' be two connections which satisfy the hypothesis of the theorem. Then $S_X - S'_X$ is a section of $E^\perp$. On the other hand, we have

$$\nabla(D^s R) = \nabla'(D^s R) = 0$$

for all s. This follows from $\nabla S = \nabla R_\nabla = 0$ and $\nabla' S' = \nabla' R_{\nabla'} = 0$ and straigfhtforward computations. Therefore,

$$0 = \nabla_X D^s R - \nabla'_X D^s R = (S_X - S'_X)\cdot D^s R$$

for all s. So, $S_X - S'_X$ is also a section of E. Hence it is zero and $\nabla = \nabla'$.

In order to prove the *existence* we consider, as in the proof of Lemma 2.1, a linear metric connection $\overline{\nabla}$ such that $\overline{\nabla} D^s R = 0$ for $0 \leq s \leq k_M + 1$. Let $\overline{S}$ be the difference tensor field between $\overline{\nabla}$ and D. $\overline{S}$ decomposes uniquely as follows

$$\overline{S} = S + \widetilde{S}, \tag{2.1}$$

where S_X is a section of $E^\perp$ and $\widetilde{S}_X$ one of E for all vector fields X on M. Let ∇ be the connection defined by

$$\nabla_X = D_X + S_X = \overline{\nabla}_X - \widetilde{S}_X. \tag{2.2}$$

Then, $\nabla_X D^s R = 0$ for $s \leq k_M + 1$. In fact,

$$\nabla_X D^s R = \overline{\nabla}_X D^s R - \widetilde{S}_X \cdot D^s R = 0$$

for $s \leq k_M + 1$, because $\widetilde{S}_X$ is a section of E. Therefore,

$$i_X D^{s+1} R = - S_X \cdot D^s R \tag{2.3}$$

for all $s \leq k_M$. Thus, $D^{s+1}R$ is obtainable from S and $D^s R$ by tensor products, contractions and suitable permutations of the arguments. Since ∇ acts on the tensor algebra as a derivation, commuting with the contractions and with the permutations of the arguments, we get

$$\nabla_X D^{s+1} R = -(\nabla_X S)\cdot D^s R$$

for $0 \leq s \leq k_M$. Therefore, $(\nabla_X S)_Y \cdot D^s R = 0$ if $0 \leq s \leq k_M$, and $(\nabla_X S)_Y$ is a section of E for all vector fields X,Y on M.

On the other hand, we have

$$(\nabla_X S)_Y = [\nabla_X, S_Y] - S_{\nabla_X Y}.$$

$S_{\nabla_X Y}$ is a section of $E^{\perp}$. Moreover, the covariant derivative of the section $\sigma = S_Y$ with respect to X is given by

$$(\nabla_X \sigma)(Z) = \nabla_X \sigma(Z) - \sigma(\nabla_X Z) = [\nabla_X, \sigma](Z) = [\nabla_X, S_Y](Z).$$

So, from the Lemma 2.2 we get that $[\nabla_X, S_Y]$ is still a section of $E^{\perp}$. Therefore, $(\nabla_X S)_Y$ is also a section of $E^{\perp}$. For this reason we have

$$\nabla S = 0.$$

Now, it is a routine matter to prove that the torsion and the curvature of ∇ are parallel with respect to ∇. This achieves the proof of Theorem 1.1.

It is also clear from the previous argument that the connection ∇ is uniquely determined by the fact that S_X is a section of $E^{\perp}$ and by (2.3). Therefore also Theorem 1.2 is proved.

Q.E.D.

We end with the following remark :

Remark 2.1. The connection constructed in Theorem 1.1 is *invariant for local isometries* of (M,g). Indeed, let f be a local isometry of M. We put

$$\tilde{S}_X = f_*^{-1} \circ S_{f_* X} \circ f_*$$

for each vector field X. Since f is a local isometry, $\tilde{S}_X$ is still a section of $E^{\perp}$.
Moreover,

$$\tilde{S}_X \cdot D^s R = S_X \cdot D^s R$$

for all $s \geq 0$, because the tensor fields $D^s R$ are invariant by local isometry and (1.5) holds for all s, since the torsion and the curvature of ∇ are parallel.

Therefore $\tilde{S}_X - S_X$ is a section of E and so $\tilde{S}_X = S_X$ for all X. From this we get immediately that f is an *affine transformation* of ∇.

References

[GR] M. GROMOV, *Partial differential relations*, Ergebnisse der Mathematik und ihrer Grenzgebiete, 3. Folge, Band 9, Springer Verlag, Berlin, Heidelberg, New York (1986).

[LA] F. LASTARIA, Homogeneous metrics with the same curvature, *Simon Stevin* (to appear).

[LT] F. LASTARIA and F. TRICERRI, Curvature-orbits and locally homogeneous Riemannian manifolds, *Ann. di Mat. pura ed appl.* (to appear).

[NT] L. NICOLODI and F. TRICERRI, On two theorems of I.M.Singer about homogeneous spaces, *Ann. Global Anal. Geom.* **8** (1990), 193-209.

[SI] I.M. SINGER, Infinitesimally homogeneous spaces, *Comm. Pure Appl. Math.* **13** (1960), 685-697.

This paper is in final form and no version will appear elsewhere.

Charles University, Faculty of Mathematics and Physics,
Sokolovska 83, 18600 Praha, Czechoslovakia.

Dipartimento di Matematica, Università di Firenze,
viale Morgagni 67/A, 50134 Firenze, Italy.

SOME IMPROPER AFFINE SPHERES IN A_3

Michael Kozlowski

A wide class of affine surfaces in A_3 is given by improper affine spheres which are surfaces with a constant equiaffine normal. This class forms a subclass of the affine maximal surfaces, i.e. their equiaffine mean curvature vanishes identically.

Suppose Ω is a region in the plane and that an affine surface $\Sigma : \Omega \longrightarrow A_3$ is given by a differentiable function z as a graph over Ω

$$\Sigma(x,y) = (x,y,z(x,y)). \tag{1}$$

Denote by d the determinant of the Hessian of z

$$d = z_{xx}z_{yy} - (z_{xy})^2. \tag{2}$$

$\Sigma : \Omega \longrightarrow A_3$ is an affine maximal surface (cf. [CAL],[SCHN],[SI]) if $z : \Omega \longrightarrow R$ satisfies the Euler-Lagrange equation

$$d\{z_{xx}d_{yy} + z_{yy}d_{xx} - 2z_{xy}d_{xy}\} = \frac{7}{4}\{z_{xx}d_y^2 + z_{yy}d_x^2 - 2z_{xy}d_xd_y\}. \tag{3}$$

$\Sigma : \Omega \longrightarrow A_3$ is an improper affine sphere with affine normal parallel to the third axis if $z : \Omega \longrightarrow R$ satisfies the Monge-Ampère equation

$$z_{xx}z_{yy} - (z_{xy})^2 = 1. \tag{4}$$

We assume that $\Sigma : \Omega \longrightarrow A_3$ only consists of elliptic points, i.e. d is positive everywhere in Ω.

THEOREM. Suppose Γ is a region in the complex plane and $f : \Gamma \longrightarrow C$ is a holomorphic function in Γ of the form

$$f(a+ib) = \mu(a,b) + i\tau(a,b)$$

where $\mu, \tau : \Gamma \longrightarrow R$ are real functions. Suppose $(a_0, b_0) \in \Gamma$ and $\tau_b(a_0,b_0) \neq O$. Then from f we can construct a solution of the Monge-Ampère equation (4).

The basic idea in the following proof is from E. Heinz and was used by Jörgens in [JÖ].

Proof. Define coordinates (x,y) in the plane by

$$x = a, \quad y = \tau(a,b).$$

Because $\tau_b(a_0,b_0) \neq O$ there exists a function $B : U(x_0,y_0) \longrightarrow \Gamma$ in a neighborhood of a point (x_0,y_0), $(x_0,y_0) = (a_0, \tau(a_0,b_0))$, such that

$$a = x, \quad b = B(x,y).$$

Define $z : U(x_0,y_0) \longrightarrow R$ by the differential

$$dz = \mu(x, B(x,y))dx + B(x,y)dy. \tag{5}$$

(i) (5) is a total differential: Setting $M := \mu(x, B(x,y))$ one has

$$M_y = \mu_b B_y = -\tau_a B_y.$$

Otherwise $b = B(x,y) = B(a, \tau(a,b))$ and differentiation by a gives

$$O = B_x + B_y \tau_a. \qquad (*)$$

(ii) z is a solution of the Monge-Ampère equation (4): Differentiation by b gives

$$1 = B_y \tau_b. \qquad (**)$$

Using Cauchy Riemann equations one has for the Hessian of z

$$\begin{aligned} d &= M_x B_y - M_y^2 = M_x B_y - \tau_a^2 B_y^2 \\ &= (\mu_a + \mu_b B_x) B_y - \tau_a^2 B_y^2 \\ &= (\tau_b - \tau_a B_x) B_y - \tau_a^2 B_y^2 \\ &\overset{(*)}{=} (\tau_b + \tau_a^2 B_y) B_y - \tau_a^2 B_y^2 \\ &= \tau_b B_y \overset{(**)}{=} 1. \end{aligned}$$

Examples
1) The complex identy map $f(w) = w$ induces the elliptic paraboloid (Figure 1)

$$z = \frac{x^2 + y^2}{2}.$$

2) A nontrivial example is obtained by the holomorphic function $f(w) = \log(\sin w)$. Then $B(x,y)$ is given by

$$B(x,y) = \operatorname{artanh}(\tan x \tan y)$$

and dz becomes .

$$dz = \frac{1}{2} \ln(\frac{1}{2}(\cosh(2B(x,y)) - \cos 2x)) dx + B(x,y) dy,$$

see Figure 2.

We introduce functions $\delta, \mu : \Omega \longrightarrow R$ by

$$\delta := d^{\frac{1}{4}}, \quad \mu := \frac{1}{\delta}. \qquad (6)$$

Then the Berwald-Blaschke metric and its inverse are given by

$$G = \mu Hess(z) = \mu \begin{pmatrix} z_{xx} & z_{xy} \\ z_{xy} & z_{yy} \end{pmatrix}, \qquad (7)$$

$$G^{-1} = \mu^3 \begin{pmatrix} z_{yy} & -z_{xy} \\ -z_{xy} & z_{xx} \end{pmatrix}. \qquad (8)$$

One has for the conormal K of the tangential plane

$$K = \mu(-z_x, -z_y, 1). \tag{9}$$

We get for the partial derivatives of K

$$\|K, K_x, K_y\| = \mu^3 \begin{pmatrix} -z_x & -z_{xx} & -z_{xy} \\ -z_y & -z_{xy} & -z_{yy} \\ 1 & 0 & 0 \end{pmatrix} = \mu^3 d = \delta > 0. \tag{10}$$

Thus one can use K, K_x, K_y as a basis of R^3. One can calculate the Weingartenform from the equations

$$K_{xx} =^2 \Gamma^1_{11} K_x +^2 \Gamma^2_{11} K_y - B_{11} K, \tag{11.a}$$

$$K_{xy} =^2 \Gamma^1_{12} K_x +^2 \Gamma^2_{12} K_y - B_{12} K, \tag{11.b}$$

$$K_{yy} =^2 \Gamma^1_{22} K_x +^2 \Gamma^2_{22} K_y - B_{22} K. \tag{11.c}$$

A short computation then gives

$$B_{11} = -\mu_{xx}\delta - 2\mu_x\delta_x - \mu^2\alpha\delta_x - \mu^2\beta\delta_y, \tag{12.a}$$

$$B_{12} = -\mu_{xy}\delta - \mu_x\delta_y - \mu_y\delta_x - \mu^2\sigma\delta_x - \mu^2\tau\delta_y, \tag{12.b}$$

$$B_{22} = -\mu_{yy}\delta - 2\mu_y\delta_y - \mu^2\epsilon\delta_x - \mu^2\theta\delta_y, \tag{12.c}$$

where $\alpha, \beta, \sigma, \tau, \epsilon, \theta$ are the functions

$$\alpha = \mu^3\{z_{yy}z_{xxx} - z_{xy}z_{xxy}\}, \tag{12.d}$$

$$\beta = \mu^3\{-z_{xy}z_{xxx} + z_{xx}z_{xxy}\}, \tag{12.e}$$

$$\sigma = \mu^3\{z_{yy}z_{xxy} - z_{xy}z_{xyy}\}, \tag{12.f}$$

$$\tau = \mu^3\{-z_{xy}z_{xxy} + z_{xx}z_{xyy}\}, \tag{12.g}$$

$$\epsilon = \mu^3\{z_{yy}z_{xyy} - z_{xy}z_{yyy}\}, \tag{12.h}$$

$$\theta = \mu^3\{-z_{xy}z_{xyy} + z_{xx}z_{yyy}\}. \tag{12.i}$$

Suppose $\Sigma : \Omega \longrightarrow A_3$ is an affine surface with $d = x^{-4/3}$. Then the Euler-Lagrange equation is solved. Take for example the surface with

$$z = -\frac{9}{2}x^{\frac{2}{3}} + \frac{1}{2}y^2, \tag{13}$$

see Figure 3. For this surface one has

$$\mu = x^{\frac{1}{3}}, \quad \delta = x^{-\frac{1}{3}}, \quad \alpha = -\frac{4}{3}x^{-\frac{4}{3}}, \quad \beta = 0. \tag{14}$$

Using (12) one gets that the Weingartenform B_{ij} vanishes, i.e. (13) describes an improper affine sphere.

References

[CAL] E. Calabi: Hypersurfaces with maximal affinely invariant area. Amer. J. Math. 104, 91-126 (1982).

[JÖ] K. Jörgens: Über die Lösungen der Differentialgleichung $rt - s^2 = 1$. Math. Annalen 127, 130-134 (1954).

[SCHN] R. Schneider: Zur affinen Differentialgeometrie im Großen I. Math. Z. 101, 375-406 (1967).

[SI] U. Simon: Zur Entwicklung der affinen Differentialgeometrie nach Blaschke. In W. Blaschke: Gesammelte Werke, vol. 4. Thales Verlag, Essen, 1985, p. 35-88.

This paper is in final form and no version will appear elsewhere.

Michael Kozlowski
Fachbereich Mathematik der Technischen Universität Berlin
Straße des 17. Juni 135, W-1000 Berlin 12

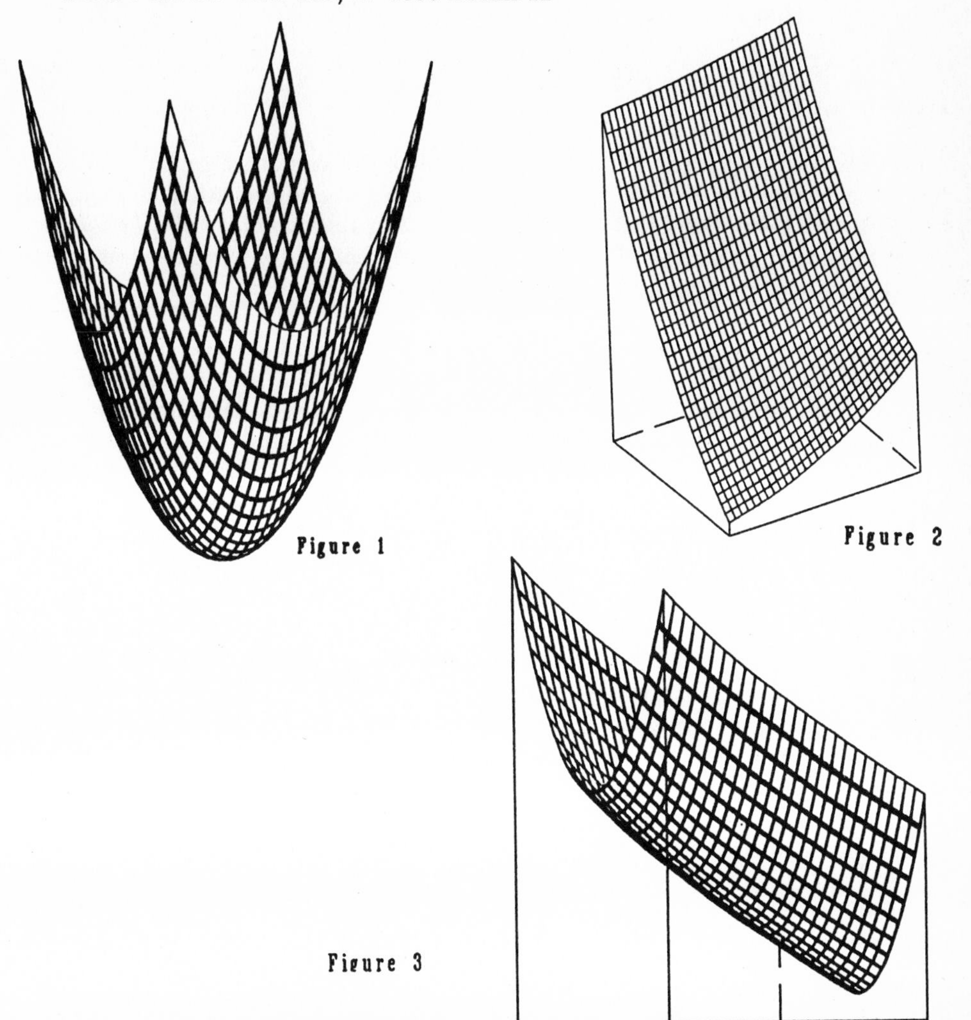

Figure 1

Figure 2

Figure 3

A Maximum Principle at Infinity and the Topology of Complete Embedded Surfaces with Constant Mean Curvature

Rob Kusner*

1 Introduction

Many new examples of complete embedded constant mean curvature surfaces $\Sigma \subset \mathbf{R}^3$ have been constructed by N. Kapouleas [4] using spheres and segments of Delaunay surfaces as the basic building blocks. Each example Σ is the boundary of a *handlebody* — a closed regular neighborhood $\overline{\Omega} \subset \mathbf{R}^3$ of a properly embedded 1-complex or *graph.* In this note we show that this simple topological behavior is typical.

Theorem 1.1 *Let $\Sigma \subset \mathbf{R}^3$ be a complete, properly embedded surface, with nonzero constant mean curvature H and finite topological type. Let $\Omega \subset \mathbf{R}^3$ be the domain of $\mathbf{R}^3 - \Sigma$ into which the mean curvature vector of Σ points. Then $\overline{\Omega}$ is a handlebody.*

We refer to Ω as the *interior* of Σ. Thus Theorem 1.1 asserts that Σ is "unknotted on the interior". We should mention here that W. Meeks and S.T. Yau [11] have recently proven that finite topology minimal ($H = 0$) surfaces are unknotted. We show this need no longer be true when $H \neq 0$ by constructing an example (Section 4) of a properly embedded constant mean curvature torus $\Sigma \subset \mathbf{R}^3$ with 7 ends which *is* "knotted on the exterior".

The proof of Theorem 1.1 (Section 3) relies on a simple adaptation of the Frankel-Lawson [3, 10] method for studying the fundamental groups of minimal surfaces in the three sphere $\mathbf{S}^3$. It also depends on an elementary comparison of mean curvatures which prohibits one annular end of Σ from lying interior to another. (For the definition of the interior of an annular end, see Section 2.) In fact, this analysis of the annular ends can be refined to yield the following "comparison principle at infinity" for constant mean curvature surfaces, answering a question of R.S. Earp and H. Rosenberg [2]:

Proposition 1.1 *Suppose A_1 and A_2 are properly embedded annular ends with positive constant mean curvatures H_1 and H_2. If A_1 is contained in the interior of A_2, then $H_1 > H_2$.*

*The research in this paper was supported by National Science Foundation grant DMS-8908064.

Its proof uses a recent result of N. Korevaar, R. Kusner, and B. Solomon [6], which asserts that *a properly embedded annular end $A \subset \mathbb{R}^3$ with nonzero constant mean curvature must converge to an end of a Delaunay surface of revolution*, together with some analysis of the Jacobi equation on a Delaunay surface. An immediate consequence of Proposition 1.1 and the result in [6] is this "maximum principle at infinity" for constant mean curvature surfaces.

Theorem 1.2 *If Σ_1 and Σ_2 are disjoint properly embedded surfaces in $\mathbf{R}^3$ with compact boundaries $\partial\Sigma_1$ and $\partial\Sigma_2$, finite topological type, and nonzero constant mean curvature H, then the distance*

$$d(\Sigma_1, \Sigma_2) \equiv \inf_{x_1 \in \Sigma_1, x_2 \in \Sigma_2} |x_1 - x_2| = 0 \quad \textit{or} \quad d(\Sigma_1, \Sigma_2) = \min_{x_1 \in \Sigma_1, x_2 \in \Sigma_2} |x_1 - x_2| > 0 .$$

The only way the first alternative holds is if Σ_1 and Σ_2 are asymptotic to Delaunay ends with disjoint interiors and parallel axes.

In the case of minimal surfaces, a maximum principle of this type was first observed by R. Langevin and H. Rosenberg [9], but the first alternative does not arise. Interestingly, the first alternative *can* occur for nonzero constant mean curvature surfaces (cf. Section 5).

2 A maximum principle at infinity

Our strategy will be to reduce Proposition 1.1 to a "vanishing lemma" (Lemma 2.1) for bounded nonnegative solutions of the Jacobi equation on a Delaunay end.

Configuration. The *Jacobi operator* on a constant mean curvature surface $\Sigma \subset \mathbb{R}^3$ is given by

$$\mathcal{J}u = \Delta u + |A|^2 u$$

where Δ is the Laplacian and $|A|$ is the length of the second fundamental form of Σ. We call u a *Jacobi field* on Σ. A *Delaunay surface* $D = D(m) \subset \mathbb{R}^3$ with mean curvature $H = (\kappa_1 + \kappa_2)/2$ and mass m (see [6] (3.9) and note that the Kapouleas "τ parameter" [4] (A.3.1) and the mass are related by $m = 2\pi\tau$) is—up to a rigid motion—the surface of revolution about the x axis defined by the ordinary differential equation

$$m = 2\pi\rho(x)\left[1/\sqrt{1 + (\rho'(x))^2} - H\rho(x)\right] \qquad \rho'(0) = 0$$

for the distance $\rho(x)$ from the x axis. This function $\rho(x)$ is bounded and periodic in x. A *Delaunay end* D^+ is the annular end $D \cap \{x \geq 0\}$.

Lemma 2.1 *Every bounded, nonnegative Jacobi field u on a Delaunay end D^+ must vanish identically* ($u \equiv 0$).

Proof. Recall from [6] (5.17) that we can solve the Jacobi equation on a Delaunay end by separating the x and θ variables:

$$u(x, \theta) = u_1(x, \theta) + \sum_{k=2}^{\infty} \sin(k\theta + \varphi_k)\, X_k(x) \qquad x \geq 0$$

where each φ_k is a constant, and—since the Jacobi field u is bounded—each X_k decays at the rate $C_k e^{-\lambda_k x}$ (for some positive λ_k strictly increasing with k). If $u_1 \not\equiv 0$, then by [6] (5.15) we can assume u_1 arises from varying D^+ through a nontrivial 1-parameter family of translations, and so u_1 must be periodic in x and change sign. Since $u - u_1$ is exponentially decaying, u also changes sign.

In case $u_1 \equiv 0$, observe that the first nonvanishing X_k is exponentially larger than the tail end of the series for large x, and we see from the series expansion that u must again change sign (unless $u \equiv 0$). □

Next we show how Proposition 1.1 follows from the vanishing lemma. We first need:

Definition 2.1 *Let $A \subset \mathbf{R}^3$ be an annular end with constant mean curvature $H \neq 0$. Let $B \subset \mathbf{R}^3$ be the smallest ball containing ∂A. The* interior *of A is the union of B and the domains of $\mathbf{R}^3 - (A \cup B)$ into which the mean curvature of A points.*

Proof of Proposition 1.1. Let D_1^+ and D_2^+ be the Delaunay ends to which the annular ends A_1 and A_2 are asymptotic. The smooth convergence of A_i to D_i^+ implies that the mean curvature of D_i^+ is also H_i. It also implies that if A_1 is contained in the interior of A_2, then D_1^+ is contained in the closure of the interior of D_2^+.

If D_1^+ lies in the interior of D_2^+ we see easily that $H_1 > H_2$, for example, by varying D_2^+ in its 1-parameter family of Delaunay surfaces (decreasing the mass) until it contacts D_1^+, and comparing mean curvatures as in [2]. Using the strong maximum principle, the only remaining case is when $D_1^+ = D_2^+ = D^+$ with the same mean curvatures $H_1 = H_2 = H$. Then by [6] (5.20) each A_i is expressed for large x as a radial graph over D^+ of a function $\rho_i(x, \theta)$, and

$$\rho_2 - \rho_1 = u(x, \theta) + o(u(x, \theta))$$

for some exponentially decaying Jacobi field u on D^+. Since A_1 lies interior to A_2, we have $\rho_2 > \rho_1$, and so eventually $u(x, \theta) > 0$ for all x large enough. But this contradicts Lemma 2.1, and completes the proof. □

Corollary 2.1 *Distinct ends on a complete, properly embedded $\Sigma \subset \mathsf{R}^3$, with nonzero constant mean curvature and finite topology, have annular representatives with disjoint interiors.*

The above corollary is also needed for the unknotting theorem in the next section.

Proof of Theorem 1.2. Observe that we can (by induction on the number of ends) confine our attention to a pair of annular ends $A_1 \subset \Sigma_1$ and $A_2 \subset \Sigma_2$. Let D_1^+ and D_2^+ be the Delaunay ends to which A_1 and A_2 (respectively) converge.

If D_1^+ and D_2^+ are not disjoint, then we see that $d(A_1, A_2) = 0$ if and only if D_1^+ and D_2^+ have parallel axes. By Corollary 2.1, we may assume D_1^+ and D_2^+ have disjoint interiors. This is the first alternative.

If D_1^+ and D_2^+ are disjoint

$$d(D_1^+, D_2^+) \equiv M > 0\,,$$

because Delaunay surfaces are periodic. Now the smooth convergence of A_i to D_i^+ implies that for any $\varepsilon > 0$ there is a compact $K = K_\varepsilon \subset \mathbf{R}^3$ such that

$$d(A_i - K, D_i^+ - K) \leq \varepsilon/2,$$

and so by the triangle inequality (picking $\varepsilon < M$)

$$d(A_1 - K, A_2 - K) > 0.$$

Since A_1 and A_2 are disjoint and properly embedded, this easily implies $d(A_1, A_2) > 0$ as required for the second alternative. □

3 Unknotting the interior

Since each end of a properly embedded constant mean curvature surface $\Sigma \subset \mathbf{R}^3$ converges to a Delaunay surface of revolution [6], we can distantly replace each end of Σ with a smoothly matching, nearly hemispherical cap to obtain a compact surface $S \subset \mathbf{R}^3$ with positive inward mean curvature. Let W be the interior of S. Using Corollary 2.1, observe that the interior Ω of Σ is obtained from W by gluing solid half cylinders to the hemispherical caps. Thus Theorem 1.1 is a consequence of the following proposition.

Proposition 3.1 *Let W be a smooth, bounded domain in $\mathbf{R}^3$ whose boundary $S = \partial\overline{W}$ has positive inward mean curvature. Then $\overline{W}$ is a handlebody.*

Proof. Note that if S is connected and the inclusion $S \subset \overline{W}$ induces a surjective map of the fundamental group $\pi_1(S)$ onto $\pi_1(\overline{W})$, then the Loop Theorem/Dehn's Lemma (see [10, 12]) will imply that $\overline{W}$ is a handlebody by standard arguments. Thus it will suffice to show that any arc in $\overline{W}$ with endpoints on S can be homotoped into S.

Suppose this were not the case. We could then find a nontrivial length minimizing arc $\gamma\colon [0,\ell] \to \overline{W}$ meeting S orthogonally at its endpoints. So the second variation of arclength

$$\ell''(\gamma, V) = -\kappa_V(\gamma(0)) - \kappa_V(\gamma(\ell)) + \int_\gamma |\nabla_{\gamma'} V|^2 \qquad (*)$$

(κ_V is the inward normal curvature of S in the direction V) would be nonnegative for any variation vectorfield V along γ which is tangent to S at its endpoints. This leads to a contradiction as in [3, 10]: Choose e_1 and e_2 to be an orthonormal basis for the tangent plane $T_{\gamma(0)}S$ of S at $\gamma(0)$ and extend these by parallel translation to vectorfields E_1, E_2 along γ. Then $E_1(\ell)$ and $E_2(\ell)$ also form an orthonormal basis for $T_{\gamma(\ell)}S$. Substituting E_1 and E_2 into (*) and summing gives:

$$\ell''(\gamma, E_1) + \ell''(\gamma, E_2) = -2h(\gamma(0)) - 2h(\gamma(\ell)) < 0$$

(h is the inward mean curvature function of S), so one of the ℓ'' terms is negative. This contradiction completes the proof. □

Remark 3.1 *Alternatively, if $\pi_1(S)$ did not map onto $\pi_1(\overline{W})$, we could use S with positive mean curvature as a barrier and construct a compact incompressible minimal surface $M \subset \overline{W}$ as in [13]. But a minimal surface $M \subset \overline{W} \subset \mathbf{R}^3$ cannot be compact!*

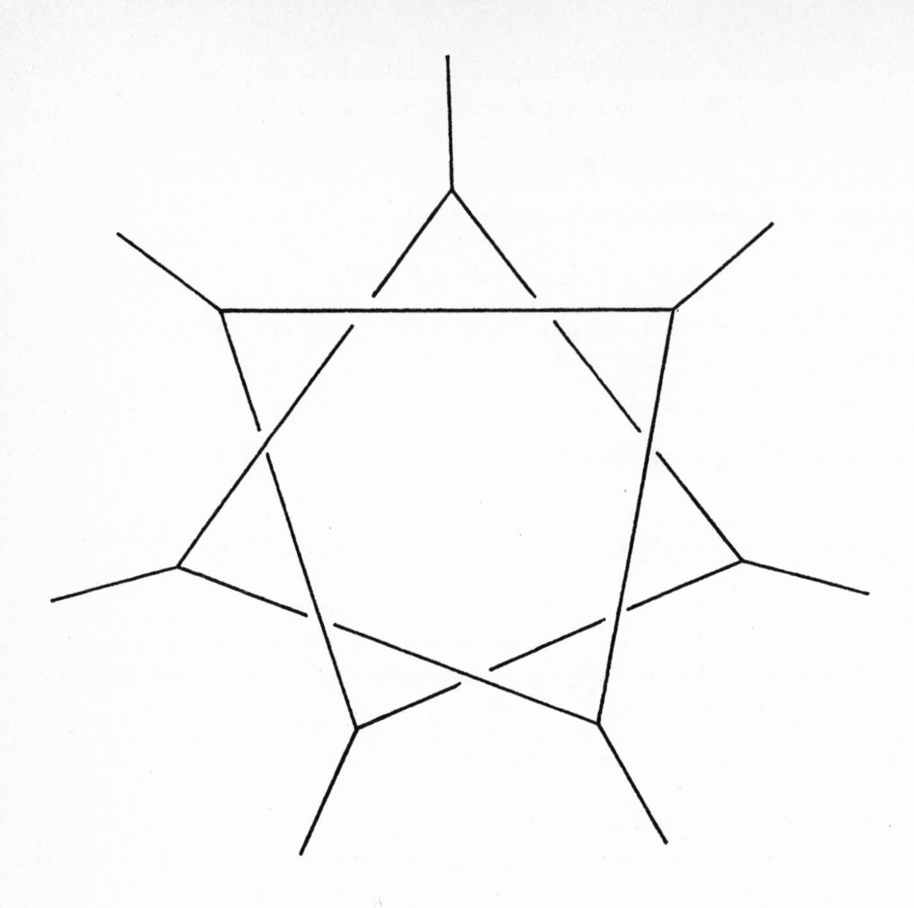

Figure 1:

4 An example with knotted exterior

In this section we construct a complete, properly embedded constant mean curvature torus $\Sigma \subset \mathbf{R}^3$ with 7 ends, whose interior handlebody $\overline{\Omega}$ is a regular neighborhood of a trefoil knot. In particular, *the exterior* $\mathbf{R}^3 - \overline{\Omega}$ *of* Σ *is knotted.*

Construction 4.1 Consider a polyhedral trefoil knot with 7 vertices which lies very close to a plane. Clearly we can arrange for the interior vertex angles to be very nearly $3\pi/7 > 60°$. Now to each vertex attach a ray so that this ray, together with the adjacent two edges, lie in a plane. Choose small positive balancing masses on the rays and edges—this is an exercise in linear algebra—to obtain a balanced graph Γ (see Figure 1).

It is easy to make Γ satisfy the conditions imposed by Kapouleas [4] (II.1) for the existence of a complete immersed constant mean curvature surface $\Sigma \subset \mathbf{R}^3$ with genus 1 and 7 ends based on Γ: in particular, Γ will be *flexible* and have even integer edge lengths. Then the almost constant mean curvature initial surface will be constructed from embedded Delaunay pieces with these positive masses, so we can assume the initial surface is embedded away from the vertices, and since each vertex angle exceeds 60°, the initial surface is embedded near the vertices as well.

Using an Alexandrov Reflection argument (cf. [5], [6] (2.13) or [4] (II.2.5)), the Kapouleas construction yields an *embedded* $\Sigma \subset \mathbf{R}^3$. The interior Ω of Σ is then isotopic to a regular neighborhood of Γ, and therefore $\mathbf{R}^3 - \overline{\Omega}$ is knotted. □

Remark 4.1 *It follows from* [6] *that any properly embedded constant mean curvature* $\Sigma \subset \mathsf{R}^3$ *with* 3 *(or fewer) ends is unknotted. Are there any knotted examples with* 4, 5, *or* 6 *ends? (Note that the preceding construction will fail to produce such examples, since—as pointed out to the author by N. Kuiper—there are* no *polyhedral knots with* 5 *or fewer vertices, and knots with* 6 *vertices must have an interior angle* $< 60°$.*)*

5 Closing Comments

The fact that the closed interior $\overline{\Omega}$ of any constant mean curvature $\Sigma \subset \mathbf{R}^3$ is a handlebody permits one to associate a *balanced diagram* to Σ as in [8]. This is a kind of "converse" to the Kapouleas construction and appears to be useful for studying the *moduli spaces* of all embedded constant mean curvature surfaces of a given genus and number of ends [5] [7].

The Jacobi field analysis of Section 2 is related to a question raised by M. do Carmo: *must the image of the Gauss map of a complete embedded constant mean curvature surface* $\Sigma \subset \mathbf{R}^3$ *always contain an equator*? By [6] this will be true unless all the asymptotically Delaunay ends of Σ converge to the ends $D^+(\frac{\pi}{2H})$ of a "pure" cylinder. However, the $\sin k\theta$ oscillation of a Jacobi field on the cylinder shows that (asymptotically) the Gauss map does *not* cover an equator. In this case, do Carmo's problem can only be solved by a global argument. A closely related question is: *suppose each end of a complete embedded surface* $\Sigma \subset \mathsf{R}^3$ *with constant mean curvature is asymptotic to a cylinder end* $D^+(\frac{\pi}{2H})$; *must* Σ be *a cylinder*? This question is very delicate because Pinkall and Sterling have recently constructed *immersed* counterexamples.

Using the methods here, in [2], and in L. Caffarelli, R. Hardt and L. Simon [1], it should be possible to extend Proposition 1.1 to: *Suppose* A *is a properly embedded annular end with constant mean curvature* H, *and* S *a surface embedded in the interior of* A *with mean curvature function* h. *Then there is a divergent sequence of points* $p_i \in S$ *such that* $h(p_i) \geq H$. These methods should also produce examples where the first alternative in Theorem 1.2 holds.

It is interesting to ask what topological properties the interior of a constant mean curvature or positive mean curvature surface S properly embedded in $\mathbf{R}^3$ enjoys if we do not assume finite topology. It seems plausible that if the mean curvature function h of S decays more slowly than the mean curvature of a parallel surface to the catenoid (with respect to intrinsic distance), then the interior is again a handlebody.

Acknowledgements. The author thanks Manfredo do Carmo, Ricardo Sá Earp and Bill Meeks for relevant discussions, Bruce Solomon for his contribution to Section 4, and Wayne Rossman for his careful reading of an earlier version of this paper.

References

[1] L. Caffarelli, R. Hardt, and L. Simon. Minimal surfaces with isolated singularities. *Manuscripta Math.*, 48:1–18, 1984.

[2] R. Earp and H. Rosenberg. Some remarks on surfaces of prescribed mean curvature. preprint.

[3] T. Frankel. On the fundamental group of a compact minimal submanifold. *Annals of Math.*, 83:68–73, 1966.

[4] N. Kapouleas. Complete constant mean curvature surfaces in Euclidean three space. *Annals of Math.*, 131:239–330, 1990.

[5] N. Korevaar and R. Kusner. The global structure of constant mean curvature surfaces. in preparation.

[6] N. Korevaar, R. Kusner, and B. Solomon. The structure of complete embedded surfaces with constant mean curvature. *Journal of Differential Geometry*, 30:465–503, 1989.

[7] R. Kusner. Compactness for families of constant mean curvature surfaces. in preparation.

[8] R. Kusner. Bubbles, conservation laws and balanced diagrams. In P. Concus, R. Finn, and D. Hoffman, editors, *Geometric Analysis and Computer Graphics*, volume 17 of *MSRI Publications*. Springer-Verlag, 1990.

[9] R. Langevin and H. Rosenberg. A maximum principle at infinity for minimal surfaces and applications. *Duke Math. Journal*, 57:819–828, 1988.

[10] H. B. Lawson. The unknottedness of minimal embeddings. *Inventiones Math.*, 11:183–187, 1970.

[11] W. H. Meeks III and S. T. Yau. The topological uniqueness of complete minimal surfaces of finite topological type. to appear in Topology.

[12] W. H. Meeks III and S. T. Yau. Topology of three-manifolds and the embedding problems in minimal surface theory. *Annals of Math.*, 112:441–484, 1980.

[13] R. Schoen and S. T. Yau. Existence of incompressible minimal surfaces and the topology of three dimensional manifolds with non-negative scalar curvature. *Annals of Math.*, 110:127–142, 1979.

Department of Mathematics, University of Massachusetts, Amherst, MA 01003

This paper is in final form and no version will appear elsewhere.

Affine Completeness and Euclidean Completeness

by Li An-Min

In affine differential geometry there are several notions of completeness, in this paper we study

1) affine completeness, that is, completeness of the Blaschke metric on M;
2) Euclidean completeness, that is, completeness of the Riemannian metric on M induced from a Euclidean metric in A^{n+1}.

S.Y. Cheng and S.T. Yau proved that for affine hyperspheres the Euclidean completeness implies the affine completeness (cf. [2], also cf. [8], [9]). In [6] we proved that every locally strongly convex, affine complete, hyperbolic affine hypersphere is Euclidean complete. In this paper we shall give a generalization of the result of S.Y. Cheng and S.T. Yau. The following theorem will be proved.

Theorem A. Let $x:M \to A^{n+1}$ be an Euclidean complete, locally strongly convex hypersurface. If there is a constant $N>0$ such that

$$(\lambda_1^2 + \lambda_2^2 + \dots + \lambda_n^2)^{1/2} \leq N ,$$

where $\lambda_1, \dots, \lambda_n$ are the affine principal curvatures, then M is affine complete.
As a consequence we will give the following partial answer of the affine Bernstein problem posed by S.S. Chern (cf. [3]):

Theorem B. Let $x_3 = f(x_1, x_2)$ be a convex function defined for all $(x_1, x_2) \in A^2$.
If $M = \{(x_1, x_2, f(x_1, x_2)) \mid (x_1, x_2) \in A^2\}$ is an affine maximal surface, and if there is a constant $N>0$ such that $|\lambda_1 \cdot \lambda_2| < N$, then M is an elliptic paraboloid.

Similar results for affine complete affine maximal surfaces were proved by A. Martinez and F. Milán (cf. [7]).

This paper was written during my stay at the FB Mathematik, TU Berlin, supported by the Alexander von Humboldt-Stiftung. I would like to thank Prof. Dr. U. Simon for discussions and for hospitality.

Let A^{n+1} be the unimodular affine space of dimension $n+1$, M be a C^∞ manifold of dimension n, and $x: M \to A^{n+1}$ a locally strongly convex hypersurface, $n \geq 2$. We choose a local unimodular affine frame field $x, e_1, \ldots, e_n, e_{n+1}$ on M such that

$$e_1, \ldots, e_n \in T_x M \quad ,$$

$$(e_1, \ldots, e_n, e_{n+1}) = 1 \quad ,$$

$$G_{ij} = \delta_{ij} \quad ,$$

$$e_{n+1} = Y \, ,$$

where G_{ij} and Y denote the Blaschke metric and the affine normal vector field resp.. Denote by A_{ijk} and B_{ij} the Fubini-Pick tensor and the affine third fundamental tensor resp.. We have the following local formulas (cf. [2], [4]):

$$x,_{ij} = \sum A_{ijk} e_k + \delta_{ij} Y \tag{1}$$

$$\Delta x = n Y \tag{2}$$

$$Y,_i = - \sum B_{ij} e_j \tag{3}$$

$$\sum A_{iik} = 0 \tag{4}$$

where `,´ denotes the covariant differentiation with respect to the Blaschke metric. The eigenvalues of (B_{ij}) are called the affine principal curvatures, and denoted by $\lambda_1, \ldots, \lambda_n$. The affine mean curvature is defined by

$$L_1 = \frac{1}{n} (\lambda_1 + \ldots + \lambda_n) \, .$$

For affine hyperspheres we have

$$\lambda_1 = \lambda_2 = \ldots = \lambda_n = \text{constant}.$$

A locally strongly convex hypersurface is called an affine maximal hypersurface if $L_1 = 0$ everywhere.

Let $x: M \to A^{n+1}$ be a non-compact, Euclidean complete, locally strongly convex hypersurface. By the Hardamard Theorem M is the graph of a strictly convex function $x_{n+1} = f(x_1, \ldots, x_n)$ defined on a convex domain $\Omega \subset A^n$. Hence M is globally strongly convex. Moreover, we may assume that the hyperplane $x_{n+1} = 0$ is the tangent hyperplane of M at some point $x_0 \in M$, and x_0 has the coordinates $(0, \ldots, 0)$. In fact, if we define

$$\tilde{f}(x_1, \ldots, x_n) = \sum \frac{\partial f}{\partial x_i}\left(\overset{\circ}{x}_1, \ldots, \overset{\circ}{x}_n\right)\left(\overset{\circ}{x}_i - x_i\right) + f(x_1, \ldots, x_n) - f(\overset{\circ}{x}_1, \ldots, \overset{\circ}{x}_n)$$

$$\tilde{x}_i = x_i - \overset{\circ}{x}_i, \quad 1 \le i \le n,$$

for any $(\overset{\circ}{x}_1, \ldots, \overset{\circ}{x}_n) \in \Omega$, then the graph of $\tilde{f}(x)$ has the required properties. Since the above transformation is affine, our claim is proved. With respect to this coordinate system we have $f \ge 0$ and for any number $C > 0$ the set

$$M_c = \{ x \in M \mid x_{n+1} = f(x_1, \ldots, x_n) \le C \}$$

is compact. Suppose that there are constants $N > 0$, $p \ge 0$ such that

$$\left(\lambda_1^2 + \lambda_2^2 + \ldots + \lambda_n^2\right)^{1/2} \le N \left[\ln(2+f) \right]^p \tag{5}$$

where $\lambda_1, \ldots, \lambda_n$ are the affine principal curvatures of M. First of all, we derive an estimate for $|\Delta f|$. Consider the function

$$\Phi = (C - f) \frac{|\Delta f|}{(2+f)[\ln(2+f)]^p} \tag{6}$$

defined on M_c. Obviously, Φ attains its supremum at some interior point x^* of M_c. Without loss of generality we may assume that $|\Delta f| \neq 0$ at x^*, then $\nabla\Phi = 0$ at x^*. Choose a local orthonormal frame field of the Blaschke metric $e_1,\dots,e_n$ on M such that, at x^*, $f_{,1} = \|\nabla f\|$, $f_{,i} = 0$ $(2 \leq i \leq n)$. Then, at x^*,

$$\frac{-f_{,1}|\Delta f|}{(2+f)[\ln(2+f)]^p} - (c-f)\frac{f_{,1}|\Delta f|}{(2+f)^2[\ln(2+f)]^p}$$

$$-p(c-f)\frac{f_{,1}|\Delta f|}{(2+f)^2[\ln(2+f)]^{p+1}} + (c-f)\frac{|\Delta f|_{,1}}{(2+f)[\ln(2+f)]^p} = 0.$$

It follows

$$\frac{f_{,1}|\Delta f|}{2+f} \leq |\Delta f|_{,1}.$$

Taking the $(n+1)$-st component of the identity

$$(\Delta x)_{,i} = nY_{,i} = -n\sum B_{ij}x_{,j}$$

we get

$$(\Delta f)_{,i} = -n\sum B_{ij}f_{,j}$$

$$|\Delta f|_{,1} \leq n(\lambda_1^2 + \dots + \lambda_n^2)^{1/2}\cdot f_{,1}.$$

Hence

$$\frac{|\Delta f| f_{,1}}{2+f} \leq nN[\ln(2+f)]^p f_{,1}.$$

In the case $x^* \neq x_0$ we have at x^* $f_{,1} = \|\nabla f\| > 0$. It follows that

$$\frac{|\Delta f|}{(2+f)[\ln(2+f)]^p} \leq nN$$

$$\Phi \leq (C-f)\,n\cdot N \leq CnN \tag{7}$$

(7) holds at x^* where Φ attains its supremum. In the case $x^*=x_0$ we have

$$\Phi \le C \left. \frac{|\Delta f|}{(2+f)[\ln(2+f)]^p} \right|_{x_0} = \frac{C}{2(\ln 2)^p} |\Delta f|(x_0)$$

Let

$$N' = \max\left\{ nN, \frac{1}{2(\ln 2)^p} |\Delta f|(x_0) \right\}.$$

Then at any point of M_c

$$\frac{|\Delta f|}{(2+f)[\ln(2+f)]^p} \le \frac{CN'}{C-f} .$$

Let $C \to \infty$, then

$$\frac{|\Delta f|}{(2+f)[\ln(2+f)]^p} \le N'. \tag{8}$$

Now we are going to derive a gradient estimate for f. Consider the function

$$\psi = \exp\left\{ \frac{-m}{C-f} \right\} \frac{\|\nabla f\|^2}{(2+f)^2[\ln(2+f)]^p} \tag{9}$$

defined on M_c, where m is a positive constant to be determined later. Clearly, ψ attains its supremum at some interior point x' of M_c. We can assume that $\|\nabla f\| > 0$ at x'. Choose a local orthonormal frame field of the Blaschke metric $e_1, \ldots, e_n$ on M such that, at x', $f_{,1} = \|\nabla f\| > 0$, $f_{,i} = 0$ $(i \ge 2)$. Then, at x',

$$\psi_{,i} = 0$$

$$\sum \psi_{,ii} \le 0 .$$

That is

$$\frac{-mf_{,i}}{(C-f)^2}\sum f_{,j}^2-\left(\frac{2}{2+f}+\frac{p}{(2+f)\ln(2+f)}\right)f_{,i}\sum f_{,j}^2+2\sum f_{,j}f_{,ji}=0 \tag{10}$$

$$-2\left[\frac{m}{(C-f)^2}+\frac{2}{2+f}+\frac{p}{(2+f)\ln(2+f)}\right]f_{,1}^2 f_{,11}$$

$$+\left[\frac{-2m}{(C-f)^3}+\frac{2}{(2+f)^2}+\frac{p(1+\ln(2+f))}{(2+f)^2[\ln(2+f)]^2}\right]f_{,1}^4$$

$$-\left[\frac{m}{(C-f)^2}+\frac{2}{2+f}+\frac{p}{(2+f)\ln(2+f)}\right](\Delta f)f_{,1}^2+2\sum f_{,ij}^2+2\sum f_{,j}f_{,jii}\leq 0 \tag{11}$$

Inserting (10) into (11) and noting

$$\frac{2}{(2+f)^2}+\frac{p(1+\ln(2+f))}{(2+f)^2[\ln(2+f)]^2}>0$$

we get

$$-\left[\frac{m}{(C-f)^2}+\frac{2}{2+f}+\frac{p}{(2+f)\ln(2+f)}\right]^2\cdot f_{,1}^4-\frac{2m}{(C-f)^3}f_{,1}^4$$

$$-\left[\frac{m}{(C-f)^2}+\frac{2}{2+f}+\frac{p}{(2+f)\ln(2+f)}\right](\Delta f)f_{,1}^2$$

$$+2\sum f_{,ij}^2+2\sum f_{,j}f_{,jii}\leq 0. \tag{12}$$

Let us now compute the terms $f_{,ij}$ and $f_{,jii}$. An application of the Ricci identity shows that

$$\sum f_{,j}f_{,jii}=\sum f_{,j}(\Delta f)_{,j}+\sum R_{ij}f_{,j}f_{,j}.$$

Since

$$(\Delta f)_{,j}=-n\sum B_{ij}f_{,i}$$

$$R_{ij} = \sum A_{mli} A_{mlj} + \frac{n-2}{2} B_{ij} + \frac{n}{2} L_1 \delta_{ij}$$

we get

$$\sum f_{,j} f_{,jii} = -nB_{11} f_{,1}^2 + \sum A_{ml1}^2 f_{,1}^2 + \frac{n-2}{2} B_{11} f_{,1}^2 + \frac{n}{2} L_1 f_{,1}^2 \tag{13}$$

Taking the (n + 1)-st component of

$$x_{,ij} = \sum A_{ijk} x_{,k} + \frac{\Delta x}{n} \delta_{ij}$$

we have

$$f_{,ij} = \sum A_{ij1} f_{,1} + \frac{\Delta f}{n} \delta_{ij},$$

$$\sum f_{,ij}^2 = \sum \left(\sum A_{ij1} f_{,1} + \frac{\Delta f}{n} \delta_{ij} \right)^2 = \sum A_{ij1}^2 f_{,1}^2 + \frac{1}{n} (\Delta f)^2 \tag{14}$$

Combination of (13) and (14) gives

$$\sum f_{,j} f_{,jii} = \sum f_{,ij}^2 + \frac{1}{2} n L_1 f_{,1}^2 - \frac{n+2}{2} B_{11} f_{,1}^2 - \frac{1}{n} (\Delta f)^2 \tag{15}$$

Applying Schwartz inequality we obtain

$$\sum f_{,ij}^2 \geq f_{,11}^2 + \sum_{i>1} f_{,ii}^2 \geq f_{,11}^2 + \frac{1}{n-1} \left(\sum_{i>1} f_{,ii} \right)^2$$

$$= \frac{n}{n-1} f_{,11}^2 + \frac{(\Delta f)^2}{n-1} - \frac{2 f_{,11} \Delta f}{n-1}$$

$$\geq \left(\frac{n}{n-1} - \delta \right) f_{,11}^2 - \frac{1 - \delta(n-1)}{\delta (n-1)^2} (\Delta f)^2 \tag{16}$$

for any $\delta > 0$. Substitute (15) and (16) into (12), we conclude from (10) that

$$\left(\frac{1}{n-1}-\delta\right)\left[\frac{m}{(C-f)^2}+\frac{2}{2+f}+\frac{p}{(2+f)\ln(2+f)}\right]^2\cdot f,_1^4-\frac{2m}{(C-f)^3}f,_1^4$$

$$-\left[\frac{m}{(c-f)^2}+\frac{2}{2+f}+\frac{p}{(2+f)\ln(2+f)}\right](\Delta f)f,_1^2+nL_1f,_1^2$$

$$-(m+2)B_{11}f,_1^2-\left[\frac{4-4(n-1)\delta}{\delta(n-1)^2}+\frac{2}{n}\right](\Delta f)^2\leq 0.$$

Let

$$\delta<\frac{1}{2(n-1)}\quad,\ m=4(n-1)C.$$

Then

$$\left(\frac{1}{n-1}-\delta\right)\left[\frac{m}{(C-f)^2}+\frac{2}{2+f}+\frac{p}{(2+f)\ln(2+f)}\right]^2\cdot f,_1^4-\frac{2m}{(C-f)^3}f,_1^4$$

$$\geq\left(\frac{1}{2(n-1)}-\delta\right)\left[\frac{m}{(C-f)^2}+\frac{2}{2+f}+\frac{p}{(2+f)\ln(2+f)}\right]^2\cdot f,_1^4\ .$$

Denote

$$g=\frac{m}{(C-f)^2}+\frac{2}{2+f}+\frac{p}{(2+f)\ln(2+f)}$$

$$a=\frac{1}{2(n-1)}-\delta>0$$

$$b=\frac{4-4(n-1)\delta}{\delta(n-1)^2}+\frac{2}{n}>0$$

We have

$$a g^2 f,_1^4-(g|\Delta f|-nL_1+(n+2)B_{11})\cdot f,_1^2-b(\Delta f)^2\leq 0.$$

It follows that

$$f,_1^2\leq\frac{1}{ag^2}\left[g|\Delta f|-nL_1+(n+2)B_{11}+g\sqrt{ab}\,|\Delta f|\right].$$

Since

$$g\geq\frac{2}{2+f}$$

$$|\Delta f| \le N'(2+f)[\ln(2+f)]^p$$

$$|-nL_1 + (n+2)B_{11}| \le [n+(n+2)](\lambda_1^2 + \dots + \lambda_n^2)^{1/2}$$

$$< 2(n+1)N[\ln(2+f)]^p$$

we get

$$f,_1^2 \le \left(\frac{N'}{2a} + \frac{\sqrt{ab}}{2a}N' + \frac{n+1}{2a}N\right)(2+f)^2[\ln(2+f)]^p$$

therefore

$$\psi \le \frac{N'}{2a} + \frac{\sqrt{ab}}{2a}N' + \frac{n+1}{2a}N$$

which holds at x', where ψ attains its suppremum. Hence at any point of M_c we have

$$\frac{|\nabla f|^2}{(2+f)^2[\ln(2+f)]^p} \le \left(\frac{N'}{2a} + \frac{\sqrt{ab}}{2a}N' + \frac{n+1}{2a}N\right)\exp\left\{\frac{4(n-1)C}{C-f}\right\}.$$

Let $C \to \infty$, then

$$\frac{|\nabla f|}{(2+f)[\ln(2+f)]^{p/2}} \le e^{2(n-1)}\sqrt{\frac{N'}{2a} + \frac{\sqrt{ab}}{2a}N' + \frac{n+1}{2a}N} := Q, \qquad (17)$$

where Q is a constant. Using the gradient estimate (17) we can prove:

Theorem 1. Let M be a locally strongly convex, Euclidean complete hypersurface in A^{n+1}. Choose a coordinate system $(x_1,\dots,x_{n+1})$ such that M can be represented by a positive strictly convex function $x_{n+1}=f(x_1,\dots,x_n)$ defined on a convex domain $\Omega \subset A^n$, and the hyperplane $x_{n+1}=0$ is the tangent hyperplane of M at some point x_0, and x_0 has coordinates $(0,\dots,0)$. If there is a constant $N>0$ such that on M

$$\left(\lambda_1^2 + \dots + \lambda_n^2\right)^{\frac{1}{2}} \le N\left[\ln(2+f)\right]^2,$$

then M is affine complete.

Proof. For any unit speed geodesic starting from x_0

$$\sigma : [0, S] \to M$$

we have

$$\frac{df}{ds} \le \| \nabla f \| \le Q(2+f)\ln(2+f) .$$

It follows that

$$S \ge \frac{1}{Q} \int_0^{x_{n+1}(\sigma(S))} \frac{df}{(2+f)\ln(2+f)} \qquad (18)$$

Since

$$\int_0^{\infty} \frac{df}{(2+f)\ln(2+f)} = \infty$$

and $f:\Omega \to R$ is proper (i.e., the inverse image of any compact set is compact), (18) implies the affine completeness of M.

(Q.E.D)

Combination of Theorem 1 and a Theorem of E. Calabi (cf. [1] Theorem 1.4, p 97) gives directly

Theorem 2. Let M be a locally strongly convex, Euclidean complete, affine maximal surface in A^3. Choose a coordinate system (x_1,x_2,x_3) such that M can be represented by a positive strictly convex function $x_3=f(x_1,x_2)$ defined on a convex domain $\Omega \subset A^2$, and the plane $x_3=0$ is the tangent plane of M at the origin. If there is a constant $N>0$ such that

$$\left(\lambda_1^2 + \lambda_2^2\right)^{1/2} \le N \left[\ln(2+f) \right]^2$$

then M is an elliptic paraboloid.

Theorem A and Theorem B are consequences of Theorem 1 and Theorem 2.

References

1. E. Calabi, Hypersurfaces with Maximal Affinely Invariant area, Amer. J. Math. 104, 91-126, (1982).

2. S.Y. Cheng and S.T. Yau, Complete Affine Hypersurfaces, Part 1 The Completeness of Affine Metrics, Comm. Pure and Appl. Math. 39 (1986) 839-866.

3. S.S. Chern, Affine Minimal Hypersurfaces, Proc. Jap-U.S. Semin. Tokyo 1977, 17-30 (1978).

4. A. Schwenk and U. Simon, Hypersurfaces with Constant Equiaffine Mean Curvature. Arch. Math. Vol. 46 (1986), 85-90.

5. Li An-Min, Calabi Conjecture on Hyperbolic Affine Hyperspheres, Math. Z. 203, 483-491 (1990).

6. Li An-Min, Calabi Conjecture on Hyperbolic Affine Hyperspheres (2), Preprint No. 248/1990, TU Berlin.

7. A. Martinez and F. Milán, On the Affine Bernstein Problem, Geom. Dedicata 37, No. 3, 295-302 (1991)

8. K. Nomizu, On completeness in affine differential geometry, Geom. Ded. 20. 43-49 (1986).

9. R. Schneider, Zur affinen Differentialgeometrie im Großen I Math. Z. 101, 375-406 (1967).

Departments of Mathematics
Sichuan University
Chengdu, Sichuan
P.R. China

On Submanifolds with parallel higher order fundamental form in Euclidean spaces

Ülo Lumiste

1. Introduction

It was proved by K. Nomizu for a Riemannian manifold M^m that $\nabla^k R = 0$ implies $\nabla R = 0$ (for the global version that requires completeness see [11]; cf. [3], Remark 7), so that the case $\nabla^k R = 0$, $\nabla^{k-1} R \neq 0$, $k > 1$ is impossible.

If M^m is immersed isometrically into a Euclidean space E^n as a submanifold, then R is determined, due to the Gauss equation, by the second fundamental form h and the Levi-Civita connection ∇ is complemented by the normal connection $\nabla^\perp$ to the van der Waerden-Bortolotti connection $\bar\nabla = \nabla \oplus \nabla^\perp$. Now the condition $\bar\nabla^p h = 0$ yields $\nabla^k R = 0$ by $k = 2p-1$ and thus $\nabla R = 0$. But can we have $\bar\nabla^{s+1} h = 0$, $\bar\nabla^s \neq 0$, $s \geq 1$, for a submanifold M^m in E^n? Does such a manifold exist? If it does, we say that it has the parallel higher order fundamental form $\bar\nabla^s h$ (of order s).

The first general results on submanifolds M^m with parallel $\bar\nabla h \neq 0$ (the case $s = 1$) were obtained by V. Mirzoyan [8,9]. The first examples of such surfaces, which give the affirmative answer to the question above for the case $m = 2$, $s = 1$, are given in [7]. The full list gives the next

Theorem 1 ([4]). A surface M^2 with parallel $\bar\nabla h \neq 0$ in a Euclidean space E^n is either
(i) a product of two lines with parallel $\bar\nabla h$, at least one of which has $\bar\nabla h \neq 0$, or
(ii) a B-scroll of a line in $S^3(r)$ with spherical curvature $k_S = as$ and with spherical torsion $\kappa_S = \pm\frac{1}{r}$, i.e. a surface $B^2(a,r)$ generated by binormal great circles of this line.

In [4] there are listed also all lines which can occur in the case (i). They are: a straight line E^1 ($h = 0$), a circle $S^1(r)$ ($\bar\nabla h = 0$), a plane clothoid $C^1(a)$ and a spherical clothoid $C^1(a,r)$ in some $S^2(r) \subset E^3$. The last two have parallel $\bar\nabla h \neq 0$; for them a is the constant ratio of the curvature (resp. geodesic curvature) and the arc length parameter.

For threedimensional submanifolds we have

Theorem 2 ([6]). Every M^3 with parallel $\bar\nabla h \neq 0$ in a Euclidean E^n is
(i) a product of three lines, which can be E^1, $S^1(r)$, $C^1(a)$ or $C^1(a,r)$ and at least one of them is a clothoid, or
(ii) a product of E^2 or $S^2(r)$ with a clothoid or
(iii) a product of $B^2(a,r)$ with one of the lines in (i).

Recall that M^m in E^n is said to be a product submanifold if $M^m = M^{m_1} \times M^{m_2}$, $0 < m_1 < m$, $M^{m_i} \subset E^{n_i}$, $E^n = E^{n_1} \times E^{n_2}$, $E^{n_1} \perp E^{n_2}$. Then the parallelity of $\bar\nabla^p h \neq 0$ for M^m is equivalent to the parallelity of $\bar\nabla^p h \neq 0$ for one of M^{m_1} or M^{m_2} and the parallelity of $\bar\nabla^q h \neq 0$, $q \leq p$, for the other. We call an M^m with parallel $\bar\nabla^p h \neq 0$ in an E^n irreducible, if it is not a such product.

The result of [5] can be now formulated as follows.

Theorem 3 ([5]). The only irreducible submanifolds M^m in an E^n with flat $\nabla^\perp$ and with parallel $\bar\nabla h \neq 0$ are $C^1(a)$, $C^1(a,r)$ and $B^2(a,r)$.

Theorem 2 shows that these $C^1(a)$, $C^1(a,r)$ and $B^2(a,r)$ are the only irreducible M^m, $m \le 3$, in an E^n with parallel $\bar{\nabla} h \neq 0$. We do not know yet, if there exists an M^m in an E^n $(m > 3)$, which has parallel $\bar{\nabla} h \neq 0$, nonflat $\nabla^\perp$ and is irreducible.

Some first general assertions on submanifolds M^m with parallel $\bar{\nabla}^s h \neq 0$, $s > 1$, are given by V. Mirzoyan in his candidate thesis (see also [8]). The full list for the next particular cases is obtained by F. Dillen [1,2].

Theorem 4 ([1]). The only hypersurfaces with parallel $\bar{\nabla}^s h \neq 0$, $s > 1$, in E^n are $C^1(pol_s) \times E^{n-2}$, where $C^1(pol_s)$ is a plane line whose curvature k is a polynomial of degree s of the arc length parameter.

Theorem 5 ([2]). If a surface M^2 with parallel $\bar{\nabla}^s h \neq 0$ in an E^n lies in a sphere $S^3(r)$ then this M^2 is a B-scroll $B^2(pol_s, r)$ of a line whose spherical curvature k_S is a polynomial of degree s of the arc length parameter and whose spherical torsion is $\kappa_S = \pm\frac{1}{r}$.

In the present paper we generalize the method of [2] to the case of higher codimension. For invariant formulations the next concepts are needed. If we have a submanifold M^m in E^n the point set

$$\{y \in T^\perp M^m \mid \overrightarrow{xy} = h(X,X),\ X \in T_x M^m \setminus \{0\}\}$$

is called the indicatrix cone of M^m at x. Adding here the restriction that X is a unit vector we get the normal curvature indicatrix of M^m at x. It is well-known that this indicatrix of an M^m with flat $\nabla^\perp$ is a simplex and the principal normal curvature vectors are radius vectors of its vertices. If in addition to it M^m has flat ∇ then these vectors are mutually orthogonal and thus the indicatrix cone is a rectangular polyhedral cone (see Section 2 below).

If U is a domain in R^m and $\phi : U \longrightarrow R$ is a smooth function then the Hessian of ϕ is the quadratic form $H_\phi(x) = f_{ij} x_i x_j$, where $f_{ij} = \frac{\partial^2 \phi}{\partial u_i \partial u_j}$. Let V^p be a Euclidean vector space and $\chi : U \longrightarrow V^p$ a smooth vector function. Then $H_\chi(x)$ is a V^p-valued quadratic form. The set of vectors $H_\chi(x)$ for a fixed $u \in U$ and for arbitrary $x \in R^m \setminus \{0\}$ is called the indicatrix cone of χ at $u \in U$.

Theorem 6. There is a bijection between the next two sets:
1) the set of congruence classes of immersions $\iota : U \longrightarrow E^n$, where U is a domain in R^m and $\iota(U)$ is a submanifold M^m with flat $\bar{\nabla}$ and parallel $\nabla^s h \neq 0$, $s \ge 1$,
2) the set of vector polynomials $\chi : U \longrightarrow V^{n-m}$ of degree $s+2$ with the property that the indicatrix cone of χ is a rectangular polyhedral cone at every point $u \in U$.

Remark that the indicatrix cone is invariant with respect to the local coordinate transformation. Thus the Theorem 6 can be generalized to the case when instead of U we have an m-dimensional smooth manifold M^m. This gives a general method (so called polynomial map method) for the investigation of submanifolds M^m with flat $\bar{\nabla}$ and parallel $\bar{\nabla}^s h \neq 0$, $s \ge 1$, in E^n. The condition of flatness of $\bar{\nabla}$ (i.e. of ∇ and $\nabla^\perp$) is not very restrictive: we know that often it follows from the parallelity of $\bar{\nabla}^s h \neq 0$.

Theorem 7. For a surface M^2 in E^n the parallelity of $\bar{\nabla}^s h \neq 0$, $s \ge 1$, implies the flatness of $\bar{\nabla}$.

Some more results in this direction are obtained recently by F. Dillen and V. Mirzoyan (private communications). The polynomial map method we demonstrate on two examples: we give a new proof of the Theorem 1 and then prove the next

Theorem 8. A surface M^2 with parallel $\bar{\nabla}^2 h \neq 0$ in an E^n is either
(i) a product of two lines or
(ii) a B-scroll $B^2(pol_2, r)$ in some $S^3(r)$ (in the sense of Theorem 5).

The new lines which we have to add to get all surfaces of the case (i) are also characterized.

2. Preliminaries.

If we have an M^m in E^n we can reduce the orthonormal frame bundle $\mathcal{O}(E^n)$ of E^n to the adapted subbundle $\mathcal{O}(M^m, E^n)$ so that in the formulae

$$dx = e_I \omega^I, \; de_I = e_J \omega_I^J, \; \omega_I^J + \omega_J^I = 0, \tag{2.1}$$

$$d\omega^I = \omega^J \wedge \omega_J^I, \; d\omega_I^J = \omega_I^K \wedge \omega_K^J, \; I, J = 1, \ldots, n \tag{2.2}$$

there hold

$$\begin{aligned} &\omega^\alpha = 0 \Rightarrow \omega^i \wedge \omega_i^\alpha = 0 \Rightarrow \omega_i^\alpha = h_{ij}^\alpha \omega^j \; (where \; h_{ij}^\alpha = h_{ji}^\alpha) \Rightarrow \\ &\bar{\nabla} h_{ij}^\alpha \wedge \omega^j = 0 \; (where \; \bar{\nabla} h_{ij}^\alpha = dh_{ij}^\alpha - h_{kj}^\alpha \omega_i^k - h_{ik}^\alpha \omega_j^k + h_{ij}^\beta \omega_\beta^\alpha) \Rightarrow \\ &\bar{\nabla} h_{ij}^\alpha = h_{ijk}^\alpha \omega^k \; (where \; h_{ijk}^\alpha = \; h_{ikj}^\alpha) \Rightarrow \bar{\nabla} h_{ijk}^\alpha \wedge \omega^k = \bar{\Omega} h_{ij}^\alpha. \end{aligned} \tag{2.3}$$

Here $i, j, \ldots = 1, \ldots, m$; $\alpha, \beta, \ldots = m+1, \ldots, n$,

$$\bar{\nabla} h_{ijk}^\alpha = dh_{ijk}^\alpha - h_{ljk}^\alpha \omega_i^l - h_{ilk}^\alpha \omega_j^l - h_{ijl}^\alpha \omega_k^l + h_{ijk}^\beta \omega_\beta^\alpha, \tag{2.4}$$

$$\Omega_i^k = d\omega_i^k - \omega_i^l \wedge \omega_l^k = -h_{i[p}^\alpha h_{q]}{}^k{}_\alpha \omega^p \wedge \omega^q, \tag{2.5}$$

$$\Omega_\alpha^\beta = d\omega_\alpha^\beta - \omega_\alpha^\gamma \wedge \omega_\gamma^\beta = -h_{i[p}^\alpha h_{q]}{}^i{}_\beta \omega^p \wedge \omega^q, \tag{2.6}$$

$$-\bar{\Omega} h_{ij}^\alpha = h_{kj}^\alpha \Omega_i^k + h_{ik}^\alpha \Omega_j^k - h_{ij}^\beta \Omega_\beta^\alpha. \tag{2.7}$$

This shows that ω_i^j and Ω_i^j are the connection and curvature forms, respectively, for ∇ and that ω_α^β and Ω_α^β are the same for $\nabla^\perp$; together they give $\bar{\nabla} = \nabla \oplus \nabla^\perp$.

In the sequence above the implications are verified by exterior differentiation and by the Cartan lemma, h_{ij}^α are the components of h, and h_{ijk}^α are the same for $\bar{\nabla} h$. From (2.3), (2.5) and (2.6) it follows that

$$\bar{\nabla} h_{ijk}^\alpha = h_{ijkl}^\alpha \omega^l, \tag{2.8}$$

where the h_{ijkl}^α, the components of $\bar{\nabla}^2 h$, can be nonsymmetric with respect to the indices k and l (but if $\bar{\nabla}$ is flat and thus $\Omega_i^j = \Omega_\alpha^\beta = 0$, they are symmetric). So we can continue the sequence (2.3):

$$\begin{aligned} &(2.8) \Rightarrow \bar{\nabla} h_{ijkl}^\alpha \wedge \omega^l = \bar{\Omega} h_{ijk}^\alpha \Rightarrow \bar{\nabla} h_{ijkl}^\alpha = h_{ijklp}^\alpha \omega^p \Rightarrow \ldots \\ &\ldots \Rightarrow \bar{\nabla} h_{ijk_1 \ldots k_s} \wedge \omega^{k_s} = \bar{\Omega} h_{ijk_1 \ldots k_{s-1}} \Rightarrow \\ &\bar{\nabla} h_{ijk_1 \ldots k_s} = h_{ijk_1 \ldots k_s k_{s+1}} \omega^{k_{s+1}} \Rightarrow \\ &\bar{\nabla} h_{ijk_1 \ldots k_{s+1}} \wedge \omega^{k_{s+1}} = \bar{\Omega} h_{ijk_1 \ldots k_s} \Rightarrow \ldots \end{aligned} \tag{2.9}$$

Here $h_{ijk_1\dots k_s}$ are the components of $\bar\nabla^s h$ and $\bar\Omega$ acts similarly as in (2.7). If $\bar\nabla$ is flat these components are symmetric with respect to all lower indices. The flatness of $\bar\nabla$ is equivalent, due to (2.5) and (2.6), to

$$h^\alpha_{ik}h_{jl\alpha} - h^\alpha_{il}h_{jk\alpha} = 0,\ h^\alpha_{ik}h_l{}^i{}_\beta - h^\alpha_{il}h_k{}^i{}_\beta = 0. \tag{2.10}$$

The last relations show that the matrices $\|h^\alpha_{ij}\|$ and $\|h^\beta_{ij}\|$ commute and by a suitable transformation of $\{e_1,\dots,e_m\}$ they can be simultaneously diagonalized, so that after this $h^\alpha_{ij} = 0$ if $i \neq j$. Now $k_i = h^\alpha_{ii}e_\alpha$ are the principal normal curvature vectors and the normal curvature indicatrix is the point set

$$\{y \in T^\perp_x M \mid \overrightarrow{xy} = \sum k_i(x^i)^2,\ \sum (x^i)^2 = 1\}.$$

As

$$\overrightarrow{xy} = k_m + \sum_a (k_a - k_m)(x^a)^2,$$

$0 \le (x^a)^2 \le 1$, this indicatrix is a simplex and for its vertices $y_1,\dots,y_m$ we have $\overrightarrow{xy}_i = k_i$. The first relations (2.10) show that every two different k_i and k_j are orthogonal, thus the indicatrix cone is a rectangular polyhedral cone.

On the other hand the flatness of $\bar\nabla$ is equivalent to the existence of a section in the bundle $\mathcal{O}(M^m,E^n)$, parallel with respect to $\bar\nabla$. This section is characterized by $\omega^j_i = \omega^\beta_\alpha = 0$ and is determined up to a constant orthogonal transformation in TM and $T^\perp M$. For this section M^m has an atlas with such local coordinates u^i that $\omega^i = du^i$ and thus from the sequence (2.3) and its continuation (2.9) it follows that

$$h^\alpha_{ijk} = \frac{\partial h^\alpha_{ij}}{\partial u^k},\ h^\alpha_{ijkl} = \frac{\partial^2 h^\alpha_{ij}}{\partial u^k \partial u^l},\ \dots,\ h^\alpha_{ijk_1\dots k_s} = \frac{\partial^s h^\alpha_{ij}}{\partial u^{k_1}\dots\partial u^{k_s}}.$$

The symmetry of h^α_{ijk} gives $\frac{\partial h^\alpha_{ij}}{\partial u^k} = \frac{\partial h^\alpha_{ik}}{\partial u^j}$ and shows that $h^\alpha_{ij}du^j$ is an exact differential $d\chi^\alpha_i$, so that $h^\alpha_{ij} = \frac{\partial \chi^\alpha_i}{\partial u^j}$ and now the symmetry of h^α_{ij} gives in the same way that $\chi^\alpha_i du^i = d\chi^\alpha$. Thus

$$h^\alpha_{ij} = \frac{\partial^2\chi^\alpha}{\partial u^i \partial u^j},\dots,h^\alpha_{ijk_1\dots k_s} = \frac{\partial^{s+2}\chi^\alpha}{\partial u^i\partial u^j\partial u^{k_1}\dots\partial u^{k_s}}.$$

The parallelity of $\bar\nabla^s h \neq 0$ is equivalent to $\bar\nabla^{s+1}h = 0$ and this is equivalent to $h_{ijk_1\dots k_s k_{s+1}} = 0$. If we have both: $\bar\nabla$ is flat and $\bar\nabla^s h \neq 0$ is parallel then with respect to some parallel section of $\mathcal{O}(M^m,E^n)$ we have

$$\frac{\partial^{s+3}\chi^\alpha}{\partial u^i\partial u^j\partial u^{k_1}\dots\partial u^{k_{s+1}}} = 0$$

and thus χ^α are polynomials of degree $\le s+2$ of $u^1,\dots,u^m$. As $\nabla^s h \neq 0$, at least one of $\chi^{m+1},\dots,\chi^n$ has degree $s+2$, so that $\chi = \chi^\alpha e_\alpha$ is a vector polynomial of degree $s+2$, where $e_{m+1},\dots,e_n$ form a normal part of some parallel section of $\mathcal{O}(M^m,E^n)$.

3. Proof of the Theorem 6.

Above we have shown that for every submanifold $M^m = \iota(U)$ with flat $\bar{\nabla}$ and parallel $\bar{\nabla}^s h \neq 0$ in E^n there is a vector polynomial $\chi : U \longrightarrow V^{n-m}$ of degree $s+2$ so that the indicatrix cone of χ is a rectangular polyhedral cone at every point $u \in U$.

Let us have conversely a vector polynomial $\chi : U \longrightarrow V^{n-m}$ with these properties. Then the indicatrix cone

$$\{y \in E^{n-m} | \ \overrightarrow{oy} = h_{ij}x^i x^j, \ (x^1, \ldots, x^m) \in R^m \setminus \{0\}\},$$

where $h_{ij} = \frac{\partial^2 \chi}{\partial u^i \partial u^j}$, has m edges for which the vectors

$$\frac{\partial (h_{kj}x^k x^j)}{\partial x^i} = h_{ij}x^j$$

are not linearly independent because the points of an edge of this cone are singular points. If the edges correspond to $(x_k^1, \ldots, x_k^m)$, $k = 1, \ldots, m$, then for every k there exist an $(x_l^1, \ldots, x_l^m) \in R^m \setminus \{0\}$, so that $x_l^i(h_{ij}x_k^j) = 0$. This shows that $(x_l^1, \ldots, x_l^m)$ gives also an edge point. For $(\kappa x_k^i + \lambda x_l^i)$ we have $h_{ij}(\kappa x_k^i + \lambda x_l^i)(\kappa x_k^j + \lambda x_l^j) = \kappa^2 h_{ij}x_k^i x_k^j + \lambda^2 h_{ij} x_l^i x_l^j$ and so get the 2-face between these edges etc. The Euclidean metric in R^m is determined so that (x_k^i) and (x_l^i) are orthogonal for every $k \neq l$. Taking the new frame vectors at every point $u \in R^m$ in their directions we get $h_{kl} = 0$ $(k \neq l)$ and so $h_{kl} = k_k \delta_{kl}$. The tensor $< h_{ik}, h_{jl} > - < h_{il}, h_{jk} >$ has the components $< k_i, k_j > (\delta_{ik}\delta_{jl} - \delta_{il}\delta_{jk})$ which are zero because the indicatrix polyhedral cone is rectangular. Thus $< h_{ik}, h_{jl} > - < h_{il}, h_{jk} >= 0$.

Due to the existence and uniqueness theorem of a submanifold for every initial frame there is an M^m in E^n with the parallel section in $\mathcal{O}(M^m, E^n)$ for which $h_{ij}^\alpha = \frac{\partial^2 \chi^\alpha}{\partial u^i \partial u^j}$ are the components of the second fundamental form. In fact, the parallelity of this section implies the flatness of ∇ and $\nabla^\perp$ which means that for this section $\omega_i^j = \omega_\alpha^\beta = 0$ and now the last equality shows that the Gauss equation is satisfied trivially. As $\bar{\nabla}_k h_{ij}^\alpha = \frac{\partial^3 \chi^\alpha}{\partial u^i \partial u^j \partial u^k}$ is symmetric with respect j and k, the Peterson-Mainardi-Codazzi equation is also satisfied. Finally, as χ is a polynomial vector of degree $s+2$ we have that $\bar{\nabla}^s h \neq 0$ is parallel.

4. Proof of the Theorem 7.

The first part of the proof - deduction of the flatness of ∇ from the parallelity of $\bar{\nabla}^s h \neq 0$ - follows the idea of V. Mirzoyan [10]. For the case of a surface M^2 the connection is flat iff $\Omega_1^2 = -K\omega^1 \wedge \omega^2$ is zero, where K is the Gauss curvature.

The proof is indirect. Suppose that $\bar{\nabla}^s h \neq 0$ is parallel and $\Omega_1^2 \neq 0$. Let us denote

$$S_{ij}^{(r)} = h_{ik_2 \ldots k_r}^\alpha h_{\alpha j}^{k_2 \ldots k_r};$$

obviously here is the symmetry with respect to i and j. The square of $\bar{\nabla}^{r-2}h$ is $S^{(r)} = S_i^{(r)i}$. If we apply the Laplace-Beltrami operator $\nabla_k \nabla^k = \Delta$ we get

$$\frac{1}{2}\Delta S^{(r)} = (\bar{\nabla}_k \bar{\nabla}^k h_{ik_2 \ldots k_r}^\alpha) h_\alpha^{ik_2 \ldots k_r} + S^{(r+1)}. \tag{4.1}$$

In a similar way as for (2.3) and (2.8) we can verify the sequence of implications

$$\nabla S_{ij}^{(r)} = S_{ijk}^{(r)}\omega^k \Rightarrow \nabla S_{ijk}^{(r)} \wedge \omega^k = \Omega S_{ij}^{(r)} \Rightarrow$$
$$\nabla S_{ijk}^{(r)} = S_{ijkl}^{(r)}\omega^l \Rightarrow \nabla S_{ijkl}^{(r)} \wedge \omega^l = \Omega S_{ijk}^{(r)},$$

where Ω acts similarly as (2.7). If $\bar{\nabla}^s h$ is parallel then $\bar{\nabla}_l h^\alpha_{ijk_1\dots k_s}$ is zero; this yields $S_{ijkl}^{(s)} = 0$ and thus $\Omega S_{ijk}^{(s)} = 0$. The last relations form a linear homogeneous 6/6-system with unknowns $S_{111}^{(s)}, S_{112}^{(s)}, \dots, S_{222}^{(s)}$, which has due to $\Omega_1^2 \neq 0$ nonvanishing determinant. It yields $S_{ijk}^{(s)} = 0$ and hence $\nabla S_{ij}^{(s)} = 0$. Substituting all this into (4.1) by $r = s$ we obtain $S^{(s+1)} = 0$ and thus $h^\alpha_{ijk_1\dots k_s} = 0$, but this contradicts to the assumption $\bar{\nabla}^s h \neq 0$.

The second part of the proof concerning the flatness of $\nabla^\perp$ can be given also in the indirect way as follows. The bundle $\mathcal{O}(M^2, E^n)$ of orthonormal frames has a canonical section, for which e_1 and e_2 correspond to the vertices of the normal curvature ellipse (or its degenerated form), but e_3 and e_4 go in the principal directions of this ellipse. Then

$$\|h_{ij}^3\| = \begin{pmatrix} \alpha + a & 0 \\ 0 & \alpha + a \end{pmatrix}, \|h_{ij}^4\| = \begin{pmatrix} \beta & b \\ b & \beta \end{pmatrix}, \|h_{ij}^5\| = \begin{pmatrix} \gamma & 0 \\ 0 & \gamma \end{pmatrix}, \|h_{ij}^\xi\| = 0,$$

where $\xi = 6, \dots, n$, a and b are semiaxis, $a \geq b \geq 0$ and α, β, γ are the coordinates of the centre of this ellipse. It follows now from (2.6) that

$$\Omega_3^4 = -2ab\omega^1 \wedge \omega^2, \; \Omega_3^\rho = \Omega_4^\rho = \Omega_\sigma^\rho = 0; \; \rho, \sigma = 5, \dots, n.$$

Suppose $\bar{\nabla}^s h \neq 0$ is parallel and $\Omega_3^4 \neq 0$; we know already that $\Omega_1^2 = 0$. The first assumption gives $h^\alpha_{ijk_1\dots k_{s+1}} = 0$ and from (2.9) thus $\bar{\Omega} h^\alpha_{ijk_1\dots k_s} = 0$ Here by $\alpha = \rho$ we get the trivial identity, but by $\alpha = 3$ or $\alpha = 4$ have

$$h^4_{ijk_1\dots k_s}\Omega_4^3 = 0, \;\; h^3_{ijk_1\dots k_s}\Omega_3^4 = 0,$$

thus $h^a_{ijk_1\dots k_s} = 0$; $a, b = 3, 4$. From $\bar{\nabla}(\bar{\nabla}^s h) = 0$ it follows that $\bar{\nabla} h^a_{ijk_1\dots k_s} = 0$ and due to (2.9) thus $h^b_{ijk_1\dots k_{s-1}}\Omega_b^a = 0$; hence $h^a_{ijk_1\dots k_{s-1}} = 0$. Here $\bar{\nabla} h^a_{ijk_1\dots k_{s-1}} = 0$ because $h^a_{ijk_1\dots k_s} = 0$ and thus $h^b_{ijk_1\dots k_{s-2}}\Omega_b^a = 0$; hence $h^a_{ijk_1\dots k_{s-2}} = 0$. So its goes further until we get the contradiction $h^3_{ij} = h^4_{ij} = 0$ to the assumption $\Omega_3^4 \neq 0$.

5. The polynomial map method for surfaces.

Let M^2 be a surface with parallel $\bar{\nabla}^s h \neq 0$ in E^n, $s \geq 1$. Due to the Theorem 7 this M^2 has flat $\bar{\nabla}$. With respect to the parallel section of $\mathcal{O}(M^2, E^n)$ the components of the second fundamental form are the second order derivatives of a vector polynomial χ of degree $\leq s + 2$ satisfying

$$< h_{uu}, h_{vv} > - h_{uv}^2 = 0, \; (h^\alpha_{uu} - h^\alpha_{vv})h^\beta_{uv} = (h^\beta_{uu} - h^\beta_{vv})h^\alpha_{uv}$$

and at least one of them has the degree $s+2$ (here $h_{uu} = \frac{\partial^2 \chi}{\partial u \partial u}$ etc.; see Section 2). The last conditions which can be given also in the form

$$< h_{uu}, h_{vv} > - h_{uv}^2 = 0, \; h_{uu} - h_{vv} \parallel h_{uv} \tag{5.1}$$

are equivalent to (2.10) if $m = 2$. Geometrically they mean that the indicatrix cone is a rectangular plane quadrant, but analytically they form a system of second order differential equations on the polynomial $\chi(u,v)$. They are invariant with respect to an orthogonal transformation

$$u = u' \cos\alpha - v' \sin\alpha, \; v = u' \sin\alpha + v' \cos\alpha;$$

especially

$$h_{u'v'} = -\frac{1}{2}(h_{uu} - h_{vv}) \sin 2\alpha + h_{uv} \cos 2\alpha, \tag{5.2}$$

$$\frac{1}{2}(h_{u'u'} - h_{v'v'}) = \frac{1}{2}(h_{uu} - h_{vv}) \cos 2\alpha + h_{uv} \sin 2\alpha.$$

It follows that the roles of $\frac{1}{2}(h_{uu} - h_{vv})$ and h_{uv} can be exchanged if to take $\alpha = \frac{\pi}{4}$. Both of them can not have the degree $< s$. In fact, if

$$\chi = \sum a_{kl} u^k v^l, \tag{5.3}$$

(summing by $k \geq 0, l \geq 0, 0 \leq k + l \leq s + 2$) then

$$h_{uu} = \sum (i+2)(i+1) a_{i+2,j} u^i v^j, \tag{5.4}$$

$$h_{vv} = \sum (j+2)(j+1) a_{i,j+2} u^i v^j, \tag{5.5}$$

$$h_{uv} = \sum (i+1)(j+1) a_{i+1,j+1} u^i v^j, \tag{5.6}$$

(summing by $i \geq 0, j \geq 0, 0 \leq i + j \leq s$). Here $deg\, h_{uv} < s$ yields $a_{kl} = 0$ if $k + l = s$, $k \geq 1$, $l \geq 1$; adding $deg(h_{uu} - h_{vv}) < s$ we get $a_{s+2,0} = a_{0,s+2} = 0$, but this contradicts to $deg \chi = s + 2$. So we can always assume that

$$deg(h_{uu} - h_{vv}) = s.$$

Let β_0 be the value by which $h^{\beta_0}_{uu} - h^{\beta_0}_{vv}$ has the minimal degree among all $h^{\beta}_{uu} - h^{\beta}_{vv}$. Let $h^{\beta_0}_{uu} - h^{\beta_0}_{vv}$ and $h^{\beta_0}_{uv}$ have the greatest common divisor r, so that

$$\frac{1}{2}(h^{\beta_0}_{uu} - h^{\beta_0}_{vv}) = qr, \; h^{\beta_0}_{uv} = pr.$$

Now the second condition (5.1) takes the form

$$(h_{uu} - h_{vv})p = 2h_{uv}q, \tag{5.7}$$

where p and q are the relatively prime polynomials of degree $\leq s$. The both sides of (5.7) are divisible by pq and if the quotient is $k \neq 0$ then

$$h_{uu} - h_{vv} = qk, \; h_{uv} = \frac{1}{2} pk, \tag{5.8}$$

where

$$deg\, k \leq s, \; deg\, q + deg\, k = s, \; deg\, p + deg\, k \leq s. \tag{5.9}$$

It is easy to see that (5.8) implies the second condition (5.1), thus these conditions are equivalent.

The next proposition gives two particular solution of our system.

Proposition. Let $\chi : U \longrightarrow V$ be a polynomial map of degree $s+2 \geq 3$ of a domain $U \subseteq R^2$ into a real Euclidean vector space V. Let the partial derivatives of χ satisfy (5.8) with (5.9) and let $< h_{uu}, h_{vv} >= h_{uv}^2$. Then up to an orthogonal transformation in the uv-plane we have the following.

(A) If $deg\, k = s$ then

$$\chi(u,v) = \chi_{(1)}(u) + \chi_{(2)}(v), \tag{5.10}$$

where $h_{uu} = \chi_{(1)}$ and $h_{vv} = \chi_{(2)}$ are mutually orthogonal.

(B) If $deg\, k = 0$ then

$$\chi(u,v) = [\pm\|a\|uv + Q(u)]e_3 + \frac{1}{2}a(u^2+v^2) + bu + cv + d, \tag{5.11}$$

where e_3 is a constant unit vector, the constant vectors $a \neq 0, b, c, d$ are orthogonal to e_3 and $Q(u)$ is a polynomial of degree $s+2$.

Conversely, (5.10) and (5.11) satisfy (5.1) without any restrictions on $deg\, k$.

To prove the assertion (B) we need a result of F. Dillen [2]: If P is a polynomial of two variables u and v of the degree $s \geq 3$ and if $H(P) = P_{uu}P_{vv} - P_{uv}^2$ is a constant then, up to an orthogonal transformation un the uv-plane, $P(u,v) = \lambda uv + Q(u)$, where $\lambda = const \neq 0$.

Proof of the Proposition. (A) Let $deg\, k = s$. Then $deg\, q = deg\, p = 0$ due to (5.9), thus p and q are constants. Now (5.2) and (5.8) give

$$h_{u'v'} = \frac{1}{2}(-q\sin 2\alpha + p\cos 2\alpha)k$$

and α can be taken so that $h_{u'v'} = 0$. After that, returning to the previous notations u and v, we have $h(u,v) = \chi_{(1)}(u) + \chi_{(2)}(v)$ and the first condition (5.1) gives

$$< \ddot{\chi}_{(1)}, \ddot{\chi}_{(2)} >= 0$$

(B) Let $deg\, k = 0$. We can take the orthonormal base $\{e_3, \ldots, e_n\}$ in V so that $e_3 \parallel k$. Then $k^\rho = 0$ $(\rho = 4,\ldots,n)$ and $h_{uu}^\rho - h_{vv}^\rho = h_{uv}^\rho = 0$. It follows that $h(u,v)^\rho = \chi_{(1)}^\rho(u) + \chi_{(2)}^\rho(v)$ and $\ddot{\chi}_{(1)}^\rho(u) = \ddot{\chi}_{(2)}^\rho(v)$. The last equality gives $h_{uu}^\rho = h_{vv}^\rho = a^\rho = const$ and so the first condition (5.1) implies $P_{uu}P_{vv} - P_{uv}^2 = -\sum(a^\rho)^2$, where $P(u,v) =< \chi(u,v), e_3 >$. The result of F. Dillen cited above gives now

$$\chi(u,v) = [\lambda uv + Q(u)]e_3 + \frac{1}{2}\sum[a^\rho(u^2+v^2) + b^\rho u + c^\rho v + d^\rho]e_\rho.$$

The first condition (5.1) yields $\lambda = \pm\|a\|$, where $a = a^\rho e_\rho$. The last assertion of the Proposition can be verified by a straightforward computation.

Theorem 1, proved in [4] by means of the principal section of the frame bundle $\mathcal{O}(M^2, E^n)$, can be obtained now as a consequence. Let $s = 1$. Then the cases (A) and (B) of the Proposition are the only possible. It remains to describe the surfaces M^2 which correspond to (5.10) and (5.11) due to the Theorem 6. Let e_1 and e_2 be the orthogonal unit vector fields on M_2, parallel with respect to the flat ∇, and let $e_3,\ldots,e_n$ be the same for the flat $\nabla^\perp$. Then in (2.1) we have $\omega_i^j=0, \omega_\alpha^\beta=0$. It leads due to (2.2) to $d\omega^i=0$, thus locally $\omega^1=du$, $\omega^2=dv$ and $dx=e_1du+e_2dv$.

For (5.10) from (2.1) and (2.3) we get $de_1=\ddot{\chi}_{(1)}(u)du$, $de_2=\ddot{\chi}_{(2)}(v)dv$. It follows that $\frac{\partial^2 x}{\partial u \partial v}=0$, thus $x=x_{(1)}(u)+x_{(2)}(v)$, $e_i=\dot{x}_{(i)}$, $\ddot{\chi}_{(i)}=\ddot{x}_{(i)}$ etc. Here $<\dot{x}_{(1)}(u),\dot{x}_{(2)}(v)>=0$ gives by differentiation that $<\overset{(k)}{x}_{(1)},\overset{(l)}{x}_{(2)}>=0$ for evey two orders k and l. Hence M^2 is a product of two lines. For (5.11) we can take $e_4 \parallel a$, so that $a=a^4e_4$, $a^4>0$. Then

$$h_{uu}^3=Q"(u),\ h_{vv}^3=0,\ h_{uv}^3=\pm a^4,$$
$$h_{uu}^4=h_{vv}^4=a^4,\ h_{uv}^4=h_{uu}^\xi=h_{vv}^\xi=h_{uv}^\xi=0;$$

$\xi=5,\ldots,n$, hence

$$\omega_1^3=Q"(u)du\pm a^4dv,\ \omega_2^3=\pm a^4du,$$
$$\omega_1^4=a^4du,\ \omega_2^4=a^4dv,\ \omega_1^\xi=\omega_2^\xi=0.$$

Here the two first equations of the last line give $da^4\wedge du=da^4\wedge dv=0$, thus $a^4=const$ and

$$d[x+(a^4)^{-1}e_4]=e_1du+e_2dv+(a^4)^{-1}(-a^4due_1-a^4dve_2)=0.$$

This shows that M^2 lies in the sphere $S^3(r)$ whose centre has radius vector $c=x+re_4$, $r=(a^4)^{-1}$. Further

$$de_1=[Q"(u)du\pm a^4dv]e_3+a^4due_4,$$
$$de_2=\pm a^4due_3+a^4dve_4,$$
$$de_3=-[Q"(u)du\pm a^4dv]e_1-(\pm a^4)due_2,$$
$$de_4=-a^4due_1-a^4dve_2.$$

We see that the v-line is a great circle of $S^3(r)$ in the direction of e_2. For the u-line the unit vectors on the tangent, on the spherical principal normal and on the spherical binormal are, respectively, e_1, e_3, and e_2, the curvature is a linear function $Q"(u)$ of the arc length u and the torsion is $\pm a^4=\pm r^{-1}$. Hence M^2 is a B-scroll, described in the Theorem 5 by $s=1$.

6. Proof of the Theorem 8.

If $s=2$ then the assertions (A) and (B) in the Proposition correspond to the cases $deg\,k=2$ and $deg\,k=0$, respectively. Repeating the argumentation of the previous proof we see that these cases lead us to the surfaces M^2 of (i) and (ii) of the Theorem 8. The only difference is that in the case (ii) $\chi(u,v)$ has the degree 4, so that $Q"(u)$ is a quadratic function of u. It remains to show that the polynomial map χ of the degree 4 satisfying (5.8) and the first condition (5.1) can be only the map of the case (A) or (B).

First we give a general scheme. Let

$$k=\sum k_{ij}u^iv^j,\ p=\sum p_{ij}u^iv^j,\ q=\sum q_{ij}u^iv^j,$$

(summing by $i, j, i+j = l$ from 0 to the degree of the left side),

$$P_{ij} = \sum k_{ab} p_{cd}, \; Q_{ij} = \sum k_{ab} q_{cd}$$

(summing by $a, b, c, d, a+c = i, b+d = j$). Then due to (5.3) - (5.6) the conditions (5.8) reduce to

$$(i+2)(i+1)a_{i+2,j} - (j+2)(j+1)a_{i,j+2} = Q_{ij}, \; (i+1)(j+1)a_{i+1,j+1} = P_{ij}.$$

It follows that

$$h_{uu} = 2a_{20} + \sum_{k=1}^{s-2}[(\frac{1}{2k}P_{k-1,1} + Q_{k0})u^k + \sum_{i+j=k} \frac{i+1}{2j} P_{i+1,j-1}u^i v^j + \frac{1}{2k}P_{1,k-1}v^k], \tag{6.1}$$

$$h_{vv} = 2a_{02} + \sum_{k=1}^{s-2}[(\frac{1}{2k}P_{k-1,1})u^k + \sum_{i+j=k} \frac{j+1}{2i} P_{i-1,j+1}u^i v^j + (\frac{1}{2k}P_{1,k-1} - Q_{0k})v^k], \tag{6.2}$$

$$h_{uv} = \frac{1}{2}\sum_{k=0}^{s-2} \sum_{i+j=k} P_{ij}u^i v^j, \tag{6.3}$$

where

$$2(a_{20} - a_{02}) = Q_{00}, \tag{6.4}$$

$$\frac{i+1}{2j}P_{i+1,j-1} - \frac{j+1}{2i}P_{i-1,j+1} = Q_{ij} \;\; (i \geq 1, \; j \geq 1). \tag{6.5}$$

The second conditions (5.1) are now satisfied. It remains to substitute (6.1) - (6.3) into the first condition (5.1) and to equate the coefficients by similar terms.

Next we do it for $deg\, k = 1$, $deg\, q = 1$, $deg\, p \leq 1$. In this case the leading coefficients in $< h_{uu}, h_{vv} >= h_{uv}^2$ give

$$< \frac{1}{4}P_{11} + Q_{20}, P_{11} >= P_{20}^2, \tag{6.6}$$

$$< \frac{1}{4}P_{11} - Q_{02}, P_{11} >= P_{02}^2, \tag{6.7}$$

$$< \frac{1}{4}P_{11} + Q_{20}, \frac{1}{4}P_{11} - Q_{02} > + < P_{20}, P_{02} >= \frac{3}{16}P_{11}^2, \tag{6.8}$$

$$< \frac{1}{4}P_{11} + Q_{20}, P_{02} >= \frac{1}{4} < P_{11}, P_{20} >, \tag{6.9}$$

$$< \frac{1}{4}P_{11} + Q_{02}, P_{20} >= \frac{1}{4} < P_{11}, P_{02} >, \tag{6.10}$$

where

$$P_{11} = k_{10}p_{01} + k_{01}p_{10}, \; P_{20} = k_{10}p_{10}, \; P_{02} = k_{01}p_{01}, \; Q_{20} = k_{10}q_{10}, \; Q_{02} = k_{01}q_{01}.$$

From the last two relations (6.9) and (6.10) it follows $< Q_{20}, P_{02} > - < Q_{02}, P_{20} >= 0$ or

$$(p_{01}q_{10} - p_{10}q_{01}) < k_{10}, k_{01} >= 0. \tag{6.11}$$

Let $p_{10} = p_{01} = 0$. Then (6.6) - (6.10) are satisfied except (6.8) which gives the equation $q_{10}q_{01} < k_{10}, k_{01} >= 0$. From (6.5) by $i = j = 1$ it follows that $q_{10}k_{01} = -q_{01}k_{10}$ and thus $q_{10}k_{01} = q_{01}k_{10} = 0$. As $deg\, k = deg\, q = 1$, we have $q_{01} = k_{01} = 0$ or $q_{10} = k_{10} = 0$ and (6.5) - (6.10) are satisfied.

Let $p_{10}^2 + p_{01}^2 \neq 0$. Now $< k_{01}, k_{10} >= 0$ is impossible, because then (6.6) and (6.7) give $p_{01}p_{10}(k_{10}^2 - k_{01}^2) = 0$; supposing $k_{01}^2 = k_{10}^2 = \kappa^2$, from $deg\, k = 1$ it follows $\kappa \neq 0$, (6.5) yields $q_{10} = -p_{01}$, $q_{01} = p_{10}$, and (6.6) - (6.8) imply a contradiction $p_{10} = p_{01} = 0$. If we suppose $p_{01} = 0$, then $p_{10} \neq 0$, (6.6) - (6.8) reduce to $4k_{10}^2 = k_{01}^2 \neq 0$, $p_{10} - 4q_{01} = p_{10} + 2q_{01} = 0$ and this is also a contradiction.

So from (6.11) it follows that $q_{10} = \lambda p_{10}$, $q_{01} = \lambda p_{01}$, $\lambda \neq 0$. Now (6.6) gives $p_{10}k_{10} - p_{01}k_{01} = \lambda P_{11}$ and from (6.9), (6.10) we get $P_{11} = 4p_{10}p_{01} < k_{10}, k_{01} >$, which leads to $p_{10}k_{01} = p_{01}k_{10}$, $P_{11} = 2p_{10}k_{01} = 2p_{01}k_{10}$ and

$$p_{10}k_{10} - (p_{01} + 2\lambda p_{10})k_{01} = 0,$$

$$(p_{10} - 2\lambda p_{01})k_{10} - p_{01}k_{01} = 0.$$

Due to $deg\, k = 1$ the determinant here mut be 0 and this gives due to $deg\, q = 1$ that $p_{10}p_{01} \neq 0$, thus $\lambda = (p_{10}^2 - p_{01}^2)/2p_{10}p_{01}$. The direct control shows that (6.5) - (6.19) are satisfied. The result is that either

1) $p_{10} = p_{01} = 0$ and $q_{01} = k_{01} = 0$ or $q_{10} = k_{10} = 0$, or

2) $q_{10} = \lambda p_{10}$, $q_{01} = \lambda p_{01}$, $p_{10}p_{01} \neq 0$, $\lambda = (p_{10}^2 - p_{01}^2)/2p_{10}p_{01} \neq 0$, $p_{10}k_{01} = p_{01}k_{10}$, $p_{10}k_{10} - p_{01}k_{01} = 2\lambda p_{10}k_{01} = 2\lambda p_{01}k_{10}$. In the case 1) we have by u^3, v^3 and uv^2 the trivial identities, but by u^2v we get $k_{10}^2 q_{10}p_{00} = 0$. Here the possibility $p_{00} = 0$ leads to $p = 0$ and gives $h_{uv} = 0$, which corresponds to the case (5.10). The other possibilities $q_{10} = 0$ or $k_{10} = 0$ contradict to $deg\, k = deg\, q = 1$. In the case 2) by u^3 and v^3 we get

$$(p_{01}q_{00} - p_{10}p_{00})k_{10}^2 + p_{10}p_{00}k_{01}^2 = 0, \tag{6.12}$$

$$p_{01}p_{00}k_{10}^2 + (p_{10}q_{00} - p_{01}p_{00})k_{01}^2 = 0. \tag{6.13}$$

Here the determinant must be 0 and thus either $q_{00} = 0$ or

$$q_{00} = \frac{(p_{10}^2 + p_{01}^2)p_{00}}{p_{10}p_{01}} \neq 0.$$

In the first case $p_{10}p_{00}(k_{10}^2 - k_{01}^2) = p_{01}p_{00}(k_{10}^2 - k_{01}^2) = 0$. Due to $deg\, q = 1$ we have $p_{10}^2 + p_{01}^2 \neq 0$. The equality $p_{00} = 0$ contradicts to the fact that p and q are relatively prime. The equality $k_{10}^2 = k_{01}^2$ shows that the collinear vectors k_{10} and k_{01} have the same length, thus $k_{10} = \pm k_{01} \neq 0$ and $p_{10} = \pm p_{01}$ which contradicts to $\lambda \neq 0$. In the second case (6.12) and (6.13) give $p_{10}^2 k_{01}^2 + p_{01}^2 k_{10}^2 = 0$ and thus $p_{10}k_{01} = p_{01}k_{10} = 0$, which contradicts to $p_{10}p_{01} \neq 0$ and $deg\, k = 1$. Hence the proof of the Theorem 8 is finished.

Remark that each of the lines which occur in the case (i) of this theorem is characterized by the property that its curvature vector $\ddot{x} = h_{11}$ is the vector polynomial

$k_2\sigma^2 + k_1\sigma + k_0$ of degree ≤ 2 of the arc length parameter σ, where the coefficients k_2, k_1, k_0 are the parallel normal vector fields on the line and at least for one of the lines $k_2 \neq 0$. If $k_2 = 0$ along the line we have E^1, $S^1(r)$, $C^1(a)$, or $C^1(a,r)$ (see [4]).

References

[1] Dillen, F.: The classification of hypersurfaces of a Euclidean space with parallel higher order fundamental form. Math. Z. 203 (1990), 635-643.

[2] Dillen, F.: Sur les hypersurfaces parallèles d'ordre supérieur. C. r. Acad. Sci. Ser. 1, V. 311 (1990), 185-187.

[3] Kobayashi, S., Nomizu, K.: Foundations of differential geometry, Vol. I. New York - London: Intersc. Publ., 1963.

[4] Lumiste, Ü.: Small-dimensional irreducible submanifolds with parallel third fundamental form. Tartu Ülikooli Toimetised. Acta et comm. Univ. Tartuensis, No. 734 (1986), 50-62 (Russian).

[5] Lumiste, Ü.: Normally flat submanifolds with parallel third fundamental form. Proc. Estonian Acad. sci. Phys. Math., V. 38, No. 2 (1989), 129-138.

[6] Lumiste, Ü.: Threedimensional submanifolds with parallel third fundamental form in Euclidean spaces. Tartu Ülikooli Toimetised. Acta et comm. Univ. Tartuensis, No. 889 (1990), 45-56.

[7] Lumiste, Ü., Mirzoyan, V.: Submanifolds with parallel third fundamental form. Tartu Ülikooli Toimetised. Acta et comm. Univ. Tartuensis, No. 665 (1984), 42-54 (Russian).

[8] Mirzoyan, V.: Submanifolds with parallel higher order fundamental form. Preprint. Tartu, 1978 (Sov. Math. Rev., 1978, 10A542 Dep.; Russian).

[9] Mirzoyan, V.: Submanifolds with commuting normal vector field. Itogi nauki i techn. VINITI. Probl. geometrii. T. 14 (1983), 73-100 (Russian).

[10] Mirzoyan, V.: On submanifolds with parallel fundamental form α_S ($s \geq 3$). Tartu Ülikooli Toimetised. Acta et comm. Univ. Tartuensis (to appear; Russian).

[11] Nomizu, K., Ozeki, H.: A theorem on curvature tensor fields. Proc. Nat. Sci. USA, V. 48 (1962), 206-207.

This paper is in final form and no version of it has appeared or will appear elsewhere.

Department of Mathematics
University of Tartu
202400 Tartu, Estonia

CONVEX AFFINE SURFACES WITH CONSTANT AFFINE MEAN CURVATURE

A. Martínez and F. Milán[(1)]

An interesting (open) problem in Affine Differential Geometry is, (see [S1]): the classification of all affine complete, noncompact, locally strongly convex surfaces M, with constant affine mean curvature H, in the unimodular real affine 3-space A^3.

The compact case was studied by Blaschke, he could prove: "*Every ovaloid in A^3 with constant affine mean curvature is an ellipsoid*".

Blaschke's assertion holds true for affine complete, locally strongly convex surfaces with positive constant affine mean curvature in A^3, (see [B] and [S2]).

The problem for affine-maximal surfaces, that is, H = 0 on M, is called Affine Bernstein Problem (see [Ch]) and states:
"*Any locally strongly convex, affine complete, affine-maximal surface M in A^3 is an elliptic paraboloid*".

Partial solutions to this problem have been obtained with additional assumptions involving M (affine sphere, ([C1],[CY2],[J] [P]), global graph, ([C2]), or some conditions in the image of the conormal map, ([C3], [L1]), and Gauss map, ([L2]).

When H=constant<0 there are known results which characterize the hyperboloid and the surface $Q(a,2)=\{(x_1,x_2,x_3)\in A^3 \mid x_1x_2x_3=a>0,\ x_1>0,\ x_2>0,\ x_3>0\}$ as complete hyperbolic affine spheres with Pick invariant satisfying some additional assumptions, ([LP], [K]).

In this communication, we give a step in the classification of the affine complete, locally strongly convex surfaces in A^3 with constant affine mean curvature. We obtain the following result,

THEOREM .- *Let M be a locally strongly convex, affine complete surface in A^3 with constant affine mean curvature H. Denote by τ, κ and J the affine Gauss-Kronecker curvature of M, the intrinsic Gaussian curvature of the affine metric and the Pick invariant, respectively. If*

(I) $J - cB^2 \geq d$, *for some real numbers c and d,* $c > \frac{2}{5}$, *and*

(II) $3J\kappa + 2HB \geq 0$,

where $B^2 = H^2 - \tau$. *Then M is one of the following surfaces:*

i) an ellipsoid,

ii) an elliptic paraboloid,

iii) an hyperboloid,

iv) an affine image of the surface Q(a,2), a>0.

---

(1) Research partially supported by DGICYT Grant PS87-0115-C03-02

Notes:

It is known the affine egregium theorem: $\kappa = J + H$. Then we have

1.- A locally strongly convex, affine complete surface in A^3 with positive constant affine mean curvature has $\kappa \geq H > 0$ and, from Bonnet's Theorem, is compact. Then, assumptions (I) and (II), in the Theorem hold and Blaschke's result is a corollary.

2.- If M is an affine-maximal surface, then assumption (II) in Theorem holds. Thus, we obtain the following partial solution of the Affine Bernstein Problem (see [MM]):

"A locally strongly convex, affine complete, affine-maximal surface in A^3, with $\kappa + c\tau$ bounded from below by a constant, for some real number $c > \frac{2}{5}$, is an elliptic paraboloid".

In particular:

If the affine Gauss-Kronecker curvature is bounded from below, we obtain that M is an elliptic paraboloid.

3.- In the case that $H < 0$ we do not assume that M is an affine sphere (B^2 vanishes identically on M, see [S3]). However, we need to assume some growth conditions for B, (expressions (I) and (II)). Using this Theorem one can obtain the following result concerning Q(a,2):

"Let M be a locally strongly convex, affine complete surface in A^3 with H=constant<0. If the affine Gauss-Kronecker curvature is bounded from below and $3\kappa \geq 2B$, then M is an affine image of the affine sphere Q(a,2)".

Proof of the Theorem.

Let Δ be the Laplacian of the affine metric. If the affine mean curvature is constant, using the integrability conditions and the basic formulas for affine surfaces one gets (see appendix)

(F1) $$\Delta(\tfrac{1}{2}J + B^2) \geq 3J\kappa + 10JB^2 + 4HB^2 - \tfrac{1}{2}J + B^2 - 4B^4,$$

(F2) $$\Delta(\tfrac{1}{2}J + B) \geq 3J\kappa + 2HB.$$

If $H > 0$, then M is compact and, from (F2), one gets $J = 0$ and $B=0$, consequently M is an ellipsoid.

If $H \leq 0$, then, from (II), either $J = 0$ (and we have a quadric) or $\kappa \geq 0$. Assume $H \leq 0$ and $\kappa \geq 0$ on M. Then, from (I) and (F1) one gets,

$$\Delta(\tfrac{1}{2}J+B^2) \geq 3J^2 + 3JH + 10JB^2 + 4HB^2 - \tfrac{1}{2}J + B^2 - 4B^4 =$$
$$= \frac{10c-4}{1+c}(\tfrac{1}{2}J+B^2)^2 + \left(\frac{28d}{c(1+c)} + 6H-1\right)(\tfrac{1}{2}J+B^2) +$$

$$+ \left(3 - \frac{10c-4}{4(1+c)}\right)J^2 + \left(10 - \frac{10c-4}{1+c}\right)JB^2 -$$

$$- \left(4 + \frac{10c-4}{1+c}\right)B^4 - \frac{14d}{c(1+c)} J + \left(2-2H- \frac{28d}{c(1+c)}\right)B^2 \geq$$

$$\geq \frac{10c-4}{1+c} \left(\tfrac{1}{2}J+B^2\right)^2 + \left(\frac{28d}{c(1+c)} + 6H-1\right)\left(\tfrac{1}{2}J+B^2\right) +$$

$$+ \left(\frac{14}{1+c} J - \frac{14c}{1+c} B^2\right)B^2 - \frac{14d}{c(1+c)} J \geq$$

$$\geq \frac{10c-4}{1+c} \left(\tfrac{1}{2}J+B^2\right)^2 + \left(\frac{28d}{c(1+c)} +6H-1\right)\left(\tfrac{1}{2}J+B^2\right) - \frac{14d^2}{c(1+c)}$$

that is, $\Delta v \geq f(v)$, where $v = \frac{1}{2}J + B^2$ and $f:\mathbb{R} \longrightarrow \mathbb{R}$ is a polynomial of degree 2 with positive principal coefficient. Using Theorem 8 of [CY1], we conclude that $\frac{1}{2}J + B^2$ is bounded from above by a constant, and $\frac{1}{2}J + B$ must be bounded from above also.

One can supposes that M is simply connected (otherwise we may pass to the universal covering surface of M). As M is affine complete with $\kappa \geq 0$ and there is no compact affine surface in A^3 with $H \leq 0$ (see [CY2]), then M is conformally equivalent to $\mathbb{C}$.

From (II) and (F2), $\frac{1}{2}J + B$ is a bounded subharmonic function on the Riemann surface $M \equiv \mathbb{C}$ which implies that $\frac{1}{2}J + B$ is constant, and $3J\kappa + 2HB = 0$. Thus, the intrinsic Gaussian curvature κ is a nonnegative constant. As M is not compact, one gets $\kappa = 0$ on M, $J = -H$ and $HB = 0$. Therefore, if $H=0$, then $J = 0$ and M is an elliptic paraboloid and if $H < 0$, then $B=0$ and M is an affine sphere with $J = -H > 0$ that is, an affine image of the surface $Q(a,2)$, $a>0$, (see [LP]).

Appendix.

Let M be an oriented, connected and convex affine surface immersed in A^3. If ξ is the affine normal of the immersion and we denote by $\overline{\nabla}$, h and S the induced connection, the affine metric and the affine Weingarten operator associated to ξ, respectively, then the equations of the immersion are given by:

(1) $\overline{R}(X,Y)Z = h(Y,Z)SX-h(X,Z)SY$ (Gauss)

(2) $(\overline{\nabla}h)(X,Y,Z) = (\overline{\nabla}h)(Y,X,Z)$ (Codazzi)

(3) $(\overline{\nabla}S)(X,Y) = (\overline{\nabla}S)(Y,X)$ (Codazzi)

(4) $h(SX,Y) = h(X,SY)$ (Ricci)

for any X, Y, Z tangent vector fields to M, $(X,Y,Z \in TM)$, where $\overline{R}$ is the curvature tensor of $\overline{\nabla}$.

Let $\hat{\nabla}$ be the Levi-Civita connection for the affine metric h.

If we denote the difference tensor between $\bar{\nabla}$ and $\hat{\nabla}$ by K, then

$$(5) \qquad K(X,Y) = \bar{\nabla}_X Y - \hat{\nabla}_X Y, \qquad X,Y \in TM$$

and one obtains the following relations

$$(6) \qquad h(K(X,Y),Z) = -(1/2)(\bar{\nabla}h)(X,Y,Z), \qquad X,Y,Z \in TM$$

$$(7) \qquad \text{trace } K_X = 0 \qquad X \in TM$$

where $K_X Y = K(X,Y)$ for any $X,Y \in TM$.

From (1), (5) and (7) it follows that the intrinsic Gaussian curvature κ of the affine metric h is given by

$$(8) \qquad \kappa h(X,Y) = h(X,Y)H + \text{trace}(K_X K_Y), \qquad X,Y \in TM$$

where

$$H = \frac{1}{2} \text{trace} S$$

is the <u>affine mean curvature</u> of the immersion.

Let $\{E_1, E_2\}$ be an orthonormal frame with respect to the affine metric and parallel at a point $x \in M$. One writes:

$$\hat{\nabla}_{E_1} E_1 = pE_2, \quad \hat{\nabla}_{E_2} E_2 = qE_1$$

$$(9) \qquad K(E_1,E_1) = aE_1 + bE_2$$

$$SE_1 = (H+\alpha)E_1 + \beta E_2$$

for some functions p, q, a, b, α, and β defined on a neighbourhood of x, then from (2), (4), (6) and (7) one gets

$$p(x) = q(x) = 0,$$

$$(10) \qquad K(E_1,E_2) = bE_1 - aE_2, \quad K(E_2,E_2) = -aE_1 - bE_2$$

$$SE_2 = \beta E_1 + (H-\alpha)E_2.$$

From (1), (8), (9) and (10),

$$(11) \qquad \kappa = H + 2(a^2+b^2) = q_1 + p_2 - p^2 - q^2, \quad J = 2(a^2+b^2),$$

$$(12) \qquad b_1 - a_2 = -\beta - 3(pa-qb), \qquad a_1 + b_2 = -\alpha + 3(bp+qa),$$

where by $(\ )_1$ and $(\ )_2$ we denote the covariant derivatives respect to E_1 and E_2 respectively.

In the rest we suppose that <u>H is constant</u>. Then from (3), (9) and (10),

$$(13)\qquad \beta_1 - \alpha_2 = 2(\alpha b - \beta a + q\beta - p\alpha), \qquad \beta_2 + \alpha_1 = 2(\beta b + \alpha a + p\beta + q\alpha).$$

If we denote by Δ and ∇ the Laplacian and the Gradient of the affine metric h, respectively, then (making the calculations at the point $x\in M$), (11) and (12) gives

$$(14)\quad a\Delta a + b\Delta b = a(a_{11}+a_{22}) + b(b_{11}+b_{22}) = 3(a^2+b^2)\kappa + a(\beta_2-\alpha_1) - b(\beta_1+\alpha_2)$$

and

$$(15)\qquad \alpha^2+\beta^2 = |\nabla a|^2 + |\nabla b|^2 - 2b_1a_2 + 2a_1b_2.$$

From (11), (14) and (15) one has

$$(16)\quad \tfrac{1}{2}\Delta J = 3J\kappa + (\alpha^2+\beta^2) + (a_1-b_2)^2 + (b_1+a_2)^2 + 2a(\beta_2-\alpha_1) - 2b(\beta_1+\alpha_2).$$

Now, using (11) and (13)

$$(17)\qquad \alpha\Delta\alpha + \beta\Delta\beta = \alpha(\alpha_{11}+\alpha_{22}) + \beta(\beta_{11}+\beta_{22}) =$$
$$= 8(a^2+b^2)B^2 + 2HB^2 + 4\beta\alpha(b_1+a_2) + 2(\alpha^2-\beta^2)(a_1-b_2)$$

where $B^2 = (\alpha^2+\beta^2) = H^2 - \det S$, thus adding the squares in (13)

$$4(a^2+b^2)B^2 = |\nabla\alpha|^2 + |\nabla\beta|^2 - 2\beta_1\alpha_2 + 2\beta_2\alpha_1$$

and

$$(18)\qquad B^{-1}(|\nabla\alpha|^2 + |\nabla\beta|^2) - B^{-3}|\alpha\nabla\alpha + \beta\nabla\beta|^2 =$$
$$= (a^2+b^2)B + \tfrac{1}{4}B^{-1}[(\alpha_1-\beta_2)^2 + (\beta_1+\alpha_2)^2] -$$
$$- B^{-3}[\tfrac{1}{2}(\alpha^2-\beta^2)(|\nabla\alpha|^2 - |\nabla\beta|^2) + 2\alpha\beta(\alpha_1\beta_1+\alpha_2\beta_2)],$$

and from (17) and (18) one gets

$$(19)\qquad \Delta B = 9(a^2+b^2)B + B^{-1}\{4\beta\alpha(b_1+a_2) + 2(\alpha^2-\beta^2)(a_1-b_2)\} -$$
$$-B^{-3}\{\tfrac{1}{2}(\alpha^2-\beta^2)(|\nabla\alpha|^2 - |\nabla\beta|^2) + 2\alpha\beta(\alpha_1\beta_1+\alpha_2\beta_2)\} + 2HB +$$
$$+ \tfrac{1}{4}B^{-1}[(\alpha_1-\beta_2)^2 + (\beta_1+\alpha_2)^2],$$

Furthermore from (13)

$$(20)\qquad \begin{aligned} b &= \tfrac{1}{2}(\alpha^2+\beta^2)^{-1}[\alpha(\beta_1-\alpha_2) + \beta(\beta_2+\alpha_1)], \\ a &= \tfrac{1}{2}(\alpha^2+\beta^2)^{-1}[\alpha(\beta_2+\alpha_1) - \beta(\beta_1-\alpha_2)], \end{aligned}$$

and one gets

$$(21)\quad (\alpha_1-\beta_2)[a(\alpha^3-3\alpha\beta^2)+b(3\alpha^2\beta-\beta^3)] + (\beta_1+\alpha_2)[a(3\alpha^2\beta-\beta^3)-b(\alpha^3-3\alpha\beta^2)] =$$

$$= \tfrac{1}{2}(\alpha^2-\beta^2)(|\nabla\alpha|^2-|\nabla\beta|^2) + 2\alpha\beta(\alpha_1\beta_1+\alpha_2\beta_2).$$

Using (16), (19), (20) and (21), one has

$$(F2)\quad \Delta(\tfrac{1}{2}J + B) = 3J\kappa + 2HB + [B^{-1}2\beta\alpha+(b_1+a_2)]^2 + [B^{-1}(\alpha^2-\beta^2)+(a_1-b_2)]^2 +$$

$$+ \tfrac{1}{3}B^{-1}\Big\{[2^{-1/2}(\alpha_1-\beta_2)-3a2^{1/2}B]^2 + [2^{-1/2}(\beta_1+\alpha_2)-3b2^{1/2}B]^2\Big\} +$$

$$+ \tfrac{1}{6}B^{-1}\Big\{\Big[2^{-1/2}(\alpha_1-\beta_2) - 3B^{-2}2^{1/2}[a(\alpha^3-3\alpha\beta^2) + b(3\alpha^2\beta-\beta^3)]\Big]^2 +$$

$$+ \Big[2^{-1/2}(\beta_1+\alpha_2)-3B^{-2}2^{1/2}[a(3\alpha^2\beta-\beta^3)-b(\alpha^3-3\alpha\beta^2)]\Big]^2\Big\} \geq$$

$$\geq 3J\kappa + 2HB .$$

In a similar way, from (13), (16) and (17), one can gets

$$(F1)\quad \Delta(\tfrac{1}{2}J + B^2) = 3J\kappa + B^2 + 10JB^2 + 4HB^2 - 4B^4 - \tfrac{1}{2}J +$$

$$+ \Big[4\beta\alpha+(b_1+a_2)\Big]^2+\Big[2(\alpha^2-\beta^2)+(a_1-b_2)\Big]^2+\Big[a+(\beta_2-\alpha_1)\Big]^2+\Big[b-(\beta_1+\alpha_2)\Big]^2 \geq$$

$$\geq 3J\kappa + B^2 + 10JB^2 + 4HB^2 - 4B^4 - \tfrac{1}{2}J.$$

References.

[B] Blaschke, W.: Vorlesungen über Differentialgeometrie II, Affine Differentialgeometrie. Berlin J. Springer 1923

[C1] Calabi, E.: The improper affine hyperspheres of convex type and a generalization of a theorem by K. Jörgens. Mich. Math. J., 5(1958), 105-126

[C2] Calabi, E.: Hypersurfaces with maximal affinely invariant area. Amer. Jour. of Math., 104(1982), 91-126

[C3] Calabi, E.: Convex affine-maximal surfaces. Results in Math., vol. 13(1988), 199-223

[CY1] Cheng, S.Y., Yau, S.T.: Differential equations on Riemannian manifolds and their geometric applications. Comm. on Pure and Applied Math., 28(1975), 333-354

[CY2] Cheng, S.Y., Yau, S.T., Complete affine hypersurfaces, Part I. The completeness of Affine Metrics. Comm. on Pure and Applied Math., 39(1986), 839-866

[Ch] Chern, S.S., Affine minimal hypersurfaces, Minimal Submanifolds and Geodesic., Kagai Publ., Ltd. Tokyo 1978, 17-30

[J] Jörgens, K.: Über die Lösungen der Differentialgleichung rt-s^2. Math. Ann., 127(1954), 180-184

[K] Kurose, T.: Two results in the affine hypersurface theory. J. Math. Soc. Japan, vol. 41, 3(1989), 539-548

[L1] Li, A.M.: Affine maximal surfaces and harmonic functions. Lec. Notes, n. 1369(1986-87), 142-151

[L2] Li, A.M.: Some theorems in affine differential geometry. Acta Math. Sinica. To appear

[LP] Li, A.M., Penn, G.: Uniquess theorems in affine differential geometry, Part II. Results in Math., vol. 13(1988), 308-317

[MM] Martínez, A., Milán, F.: On the affine Bernstein Problem. Geom. Dedicata 37, No. 3, 295-302(1991)

[P] Pogorelov, A. V.: On the improper affine hyperspheres. Geometriae Dedicata, 1(1972), 33-46

[S1] Simon, U.: Affine differential geometry. Proceedings Conf. Math. Reasearch Institute at Oberwolfach, Nov. 2-8, 1986

[S2] Simon, U.: Hypersurfaces in equiaffine differential geometry and eigenvalue problems. Proceedings Conf. Diff. Geom. Nové Mesto(CSSR) 1983; Part I, 127-136(1984)

[S3] Simon, U.: Hypersurfaces in equiaffine differential geometry, Geometriae Dedicata, 17(1984), 157-168

DEPARTAMENTO DE GEOMETRIA Y TOPOLOGIA
UNIVERSIDAD DE GRANADA
18071 GRANADA. SPAIN.

This paper is in final form and no version will appear elsewhere.

TRANSVERSAL CURVATURE AND TAUTNESS FOR RIEMANNIAN FOLIATIONS

MAUNG MIN–OO
DEPARTMENT OF MATHEMATICS & STATISTICS
MCMASTER UNIVERSITY
HAMILTON, ONTARIO
CANADA L8S 4K1

ERNST A. RUH
DEPARTMENT OF MATHEMATICS, OHIO STATE UNIVERSITY
321 WEST 18TH AVENUE, COLUMBUS, OHIO 43210

PHILIPPE TONDEUR
DEPARTMENT OF MATHEMATICS, UNIVERSITY OF ILLINOIS
1409 WEST GREEN STREET, URBANA, ILLINOIS 61801

In recent papers, the following cohomology vanishing results where established [H] [MRT].

THEOREM 1. *Let $\mathcal{F}$ be a Riemannian foliation of codimension $q \geq 2$ on a closed oriented Riemannian manifold (M, g) with bundle–like metric g. Let $\rho : Q \to Q$ be the transversal Ricci operator on the normal bundle, and $\mathcal{R} : \Lambda^2 Q \to \Lambda^2 Q$ the transversal curvature operator. Then the following holds:*
(i) if $\rho > 0$, then $H^1_B(\mathcal{F}) = 0$;
(ii) if $\mathcal{R} > 0$, then $H^r_B(\mathcal{F}) = 0$ for $0 < r < q$.

Here $H^\bullet_B(\mathcal{F})$ denotes the basic cohomology of the foliation, defined as the cohomology of the basic complex of differential forms

$$\Omega_B(\mathcal{F}) = \{\omega | \iota_X \omega = 0, \theta_X \omega = 0 \quad \text{for all vector fields} \quad X \quad \text{tangent to} \quad \mathcal{F}\}$$

The differential d_B is the restriction of the ordinary d to the subspace $\Omega_B(\mathcal{F})$ of the ordinary DeRham complex Ω_M. The transversal curvature data are associated to the canonical metric and torsionfree connection in the normal bundle Q.

A Riemannian foliation is taut, if there exists a bundle–like metric for which all leaves are minimal submanifolds. The purpose of this note is to point out the following consequence of Theorem 1.

THEOREM 2. *Let $\mathcal{F}$ be a Riemannian foliation of codimension $q \geq 2$ as in Theorem 1. If the hypothesis of either (i) or (ii) is satisfied then $\mathcal{F}$ is taut.*

REMARK: If $\mathcal{F}$ is transversally oriented, then the transversal volume form $\nu \in \Omega^q_B(\mathcal{F})$ associated to a metric with minimal leaves gives rise to a non–trivial cohomology class $[\nu] \neq 0 \in H^q_B(\mathcal{F})$, and hence necessarily $H^q_B(\mathcal{F}) \cong \mathbb{R}$ [KT]. It is a recent result of Masa [M] that this condition is conversely sufficient to guarantee the tautness of $\mathcal{F}$.

PROOF OF THEOREM 2: According to a recent result of Alvarez López, for every Riemannian foliation there is a well-defined cohomology class $\zeta(\mathcal{F}) \in H^1_B(\mathcal{F})$, whose vanishing characterizes the tautness of $\mathcal{F}$ [A]. Under the curvature assumption of Theorem 1, the first basic cohomology group vanishes, and hence $\zeta(\mathcal{F}) = 0$.

An alternative argument is to use a recent result of Kamber and Roe, according to which every Riemannian foliation admits a bundle-like metric g for which the mean curvature form κ is a basic 1-form. $\mathcal{F}$ is thus tense in the parlance of [KT]. By [KT, (4.4)] we have then $d\kappa = 0$, and $[\kappa] \in H^1_B(\mathcal{F})$ This is then a concrete realization of the cohomology class $\zeta(\mathcal{F})$ mentioned above. Under the curvature assumptions of Theorem 1, we have necessarily $[\kappa] = 0$, and hence $\kappa = df$. Modifying g as in [KT, (4.6)] to

$$\bar{g} = (e^f)^{2/p} \cdot g_L \oplus g_Q,$$

one obtains a metric with minimal leaves. In this last formula, g_L denotes the metric along the leaves induced by g, and g_Q the induced metric on the normal bundle. The integer p is the leaf dimension of $\mathcal{F}$, with $p + q = \dim M$.

REFERENCES

[A]. J. A. Alvarez López, *The basic component of the mean curvature of Riemannian foliations*, Math Ann. (to appear).

[H]. J. Hebda, *Curvature and focal points in Riemannian foliations*, Indiana Univ. Math. J. **35** (1986), 321-331.

[KT]. F. W. Kamber and Ph. Tondeur, *Foliations and metrics*, Proc. of the 1981–82 Year in Differential Geometry, Univ. of Maryland, Birkhäuser, Progress in Math. **32** (1983), 103–152.

[M]. X. M. Masa, *Duality and minimality in Riemannian foliations*, (to appear).

[MRT]. M.Min–Oo, E. A. Ruh and Ph. Tondeur, *Vanishing theorems for the basic cohomology of Riemannian foliations*, Journal für die reine und angewandte Mathematik (to appear).

[R]. B. Reinhart, *Differential Geometry of Foliations*, Ergeb. Math. **99** (1983), Springer Verlag, New York..

[T]. Ph. Tondeur, "Foliations on Riemannian manifolds," Springer Universitext, 1988.

NOTE: This article is in final form and no version has appeared nor will appear elsewhere.

Schrödinger operators associated to a holomorphic map

SEBASTIÁN MONTIEL*
ANTONIO ROS*

Universidad de Granada
Spain

In this work we will expose certain ideas and results concerning a kind of Schrödinger operators which can be considered on a compact Riemann surface. These operators will be constructed by using as a potential the energy density of a holomorphic map from the surface to the two-sphere. Besides the interest that their study has from an analytical point of view, we will see that they appear, in a natural way, in different geometrical situations such as the study of the index of complete minimal surfaces with finite total curvature and the study of the critical points of the Willmore functional.

This paper is, in fact, an expanded version of an invited lecture given by the first author in the Global Differential Geometry and Global Analysis Conference held at the Technische Universität of Berlin in June, 1990.

INTRODUCTION AND PRELIMINARIES

Let Σ be a compact Riemann surface and $\phi : \Sigma \to \mathbf{S}^2$ a holomorphic map from this surface to the unit two-sphere $\mathbf{S}^2$. Consider any metric ds^2 on Σ compatible with the complex structure and let Δ and ∇ be its Laplacian and gradient respectively. Having chosen this metric, one has the following Schrödinger operator

$$L = \Delta + |\nabla\phi|^2. \tag{1-1}$$

Our aim here is to study spectral properties of these operators and relate them to the map ϕ and the surface Σ. Of course, the eigenvalues and eigenfunctions of such an operator L depend strongly on the metric ds^2. However, we want to obtain information from it which only refers to ϕ and Σ. This can be done in two ways: first, by looking for spectral properties of L which are independent on the chosen metric; or, second, by putting on Σ a particular metric especially related to our problem.

First, denote by Q_ϕ the quadratic form corresponding to the self–adjoint operator L, that is

$$Q_\phi(u,u) = \int_\Sigma \left\{ |\nabla u|^2 - |\nabla\phi|^2 u^2 \right\} dA \quad u \in W_1(\Sigma) \tag{1-2}$$

where dA is the measure associated to the metric ds^2 and $W_1(\Sigma)$ is the corresponding Sobolev space. Because of the conformal invariance of the Dirichlet integral, the form Q_ϕ does not depend on the chosen metric ds^2. Hence, all these Schrödinger operators L in (1-1) have the same number of bounded states, that is, the same number of negative

*Partially supported by a DGCICYT Grant No. PS87–0115

eigenvalues, and the same kernel. So, we may define *the index of the holomorphic map* ϕ as the index of the quadratic form Q_ϕ:

$$\text{(1-3)} \qquad \operatorname{Ind} \phi = \operatorname{index} Q_\phi = \# \text{ bounded states of any } L$$

and *the nullity space* $N(\phi)$ *of* ϕ as the common kernel of all these operators L:

$$\text{(1-4)} \qquad N(\phi) = \text{kernel of any } L = \left\{ u \in C^\infty(\Sigma) \,|\, \Delta u + |\nabla\phi|^2 u = 0 \right\}.$$

The dimension of this space will be called *the nullity of the holomorphic map* ϕ and we will represent it by

$$\text{(1-5)} \qquad \operatorname{Nul}(\phi) = \dim N(\phi).$$

Notice that, as ϕ is holomorphic, then ϕ is harmonic, that is

$$\Delta\phi + |\nabla\phi|^2\phi = 0.$$

So, the space $L(\phi)$ of the linear functions of the components of ϕ is contained in $N(\phi)$, that is

$$\text{(1-6)} \qquad L(\phi) = \left\{ \langle \phi, a \rangle \,|\, a \in \mathbf{R}^3 \right\} \subset N(\phi)$$

and, so, Nul $\phi \geq 3$ provided that ϕ is not a constant map. It is also clear that the index of a holomorphic map ϕ, such as we have just define it, vanishes if and only if ϕ is constant.

The second way that we have mentioned above is to choose on the surface Σ a particular metric coming from the situation that we are considering. We have, at once, a natural candidate: the metric ds_ϕ^2 induced on Σ by ϕ from the standard metric ds_0^2 of the sphere $\mathbf{S}^2$, that is

$$\text{(1-7)} \qquad ds_\phi^2 = \phi^* \, ds_0^2 = \frac{|\nabla\phi|^2}{2} ds^2.$$

This metric has constant one Gauss curvature, conical singularities at the branching points of ϕ and finite area $4\pi \deg \phi$. The Schrödinger operator L_ϕ that one gets by using that metric is nothing but

$$\text{(1-8)} \qquad L_\phi = \Delta_\phi + 2,$$

where Δ_ϕ is the Laplacian of the branched metric ds_ϕ^2. Even though the metric ds_ϕ^2 is not regular, the eigenvalues and eigenfunctions of this Laplacian are well defined via a variational approach, since the codimension of the singularities set is two (see [**Ty**]). Hence, if λ is an eigenvalue of L_ϕ, its corresponding eigenspace is given by

$$\text{(1-9)} \qquad V_\lambda(\phi) = \left\{ u \in W_1(\Sigma) \,|\, Q_\phi(u,v) = \lambda \int_\Sigma uv \, dA_\phi, \ \forall v \in W_1(\Sigma) \right\}.$$

It follows, from elliptic regularity, that $V_\lambda(\phi) \subset C^\infty(\Sigma)$.

So, the spectrum of the operator L_ϕ is, up to a constant, the spectrum of the Laplacian Δ_ϕ of the branched metric ds^2_ϕ. In fact, in terms of this metric, the index of a holomorphic map that we defined above can be interpreted as the number of eigenvalues of its Laplacian which are less than two,

$$\text{(1-10)} \qquad \operatorname{Ind} \phi = \# \text{ eigenvalues of } \Delta_\phi < 2.$$

These eigenvalues are especially important because they do not come from the spectrum of the standard sphere $\mathbf{S}^2$, since its first non zero eigenvalue is exactly two. Also, one has that

$$\text{(1-11)} \qquad \operatorname{Nul} \phi = \text{multiplicity of 2 as an eigenvalue of } \Delta_\phi.$$

It is important to remark that the spectrum of the metric ds^2_ϕ is a sequence naturally associated to the holomorphic map ϕ. So, an interesting question is: What kind of information about the complex structure of the surface Σ and the map ϕ can be recovered from that sequence?

Another particular metric that one could consider on the surface Σ in order to study the corresponding Schrödinger operator L given by (1-1) is the hyperbolic metric with constant -1 curvature, provided that the genus of the surface is greater than one. In this case, that we have not studied in detail, one can see from **[Gui–K]** and from Lemma 7 below that, in the most of the cases, the spectrum of the operator L determines the map ϕ up to an isometry of $\mathbf{S}^2$.

This paper is devoted to the study of these spectral invariants associated to a holomorphic map and is organized as follows. Before stating the results, we will deal with three geometrical topics where the invariants that we have described appear: the theory of complete minimal surfaces in $\mathbf{R}^3$, the study of the Willmore surfaces and the study of the determinant of the Laplacian of metrics on compact surfaces. So, in this way, these invariants will get different geometrical meanings.

After that, we will show some results that we have obtained recently on this subject. In fact, we will prove that each function in the nullity space of a holomorphic map ϕ, which is not a linear function of $L(\phi)$, see (1-6), can be represented as the support function of a branched complete minimal surface in $\mathbf{R}^3$ with planar ends and whose extended Gauss map is ϕ. Next, we will obtain some information about the holomorphic maps with the lowest index. Also, we will get lower and upper bounds for the index and the nullity when the branching values of the holomorphic map are in an special position on the sphere. In particular, if all these branching values lie in an equator, we will compute explicitly these invariants.

In the case that our surface Σ has genus zero, we will show that the index and the nullity of $A \circ \phi$ coincide with those of ϕ, for each Möbius transformation A of the sphere. Also, we will compute the index and the nullity of a generic ϕ and give general bounds for these invariants.

We will finish the paper by making a detailed study of the index and the nullity for a holomorphic map $\phi : \overline{\mathbf{C}} \to \mathbf{S}^2$ with degree three and we will see that its index is five and its nullity is three, except when its four ramification points form an equianharmonic quadruple, that is, they are placed, up to a Möbius transformation, on the vertices of a regular thetrahedron. In this case the index is four and the nullity five.

Some related geometrical problems

In this section, we will see that there are some geometrical problems involving a compact Riemann surface Σ where a holomorphic map ϕ from the surface to the two-sphere appears and where a good knowledge of the quadratic form Q_ϕ defined in (1-2) provides an important tool in order to solve them.

Complete minimal surfaces in $\mathbf{R}^3$ with finite total curvature. Let M be an orientable surface and $f : M \to \mathbf{R}^3$ a minimal immersion into the three-dimensional Euclidean space. Recall that such an immersion is a critical point of the area for all perturbations of M with compact support. Denote by $N : M \to \mathbf{S}^2$ its Gauss map. If $D \subset M$ is a compact domain of M and $f_t : D \to \mathbf{R}^3$ is a variation of f whose variational field is uN, where $u \in C_0^\infty(D)$, then the second derivative of the induced area $A(f_t)$ is given by

$$\left.\frac{d^2}{dt^2}\right|_{t=0} A(f_t) = \int_D \{|\nabla u|^2 + 2Ku^2\}\, dA$$

where K is the Gauss curvature function of the surface M. So, *the second variation operator L of the area* is

$$L = \Delta - 2K = \Delta + |\nabla N|^2.$$

From this second variation operator L we have, for each compact domain D of M, a measure of its instability: *the index of the domain D* defined as follows:

$$\text{index } D = \#\text{ negative eigenvalues of } L \text{ with Dirichlet boundary condition.}$$

It is a classical result due to Schwarz [**Schw**] (see also [**Ba–do C**] for a stronger formulation) that, if D is small enough, then D is stable, that is the index of D vanishes. Now, for taking arbitrary large domains on the surface, suppose that M is complete. Then, in this case, Fischer-Colbrie and Schoen [**FC–Sch**], independently do Carmo and Peng [**do C–Pe**] and, independently also, Pogorelov [**Po**] showed that a large enough piece of M is unstable, provided that M is not a plane. So, it seems natural to ask: How does the index of a compact domain of a complete minimal surface in $\mathbf{R}^3$ change when its size increases? To answer this, the index of the whole of M is defined in the following way:

$$\text{index } M = \sup_{D \subset M} \text{index } D.$$

This number can become infinite. By the way, the following nice theorem was proved by Fischer-Colbrie [**FC**] and independently by Gulliver and Lawson [**Gu**], giving a geometrical consistency to the index of a complete minimal surface defined above:

$$\text{index } M < \infty \Longleftrightarrow \left|\int_M K\, dA\right| < \infty,$$

that is, this index is finite if and only if the total Gaussian curvature of the surface is finite. Moreover, complete minimal surfaces with finite total curvature form a well-known family. In fact, through the work of Osserman, see [**Oss**], we know that such a surface has a conformal structure equivalent to that of a compact Riemann surface Σ after removing a finite number of its points. On the other hand, the Gauss map N

of this surface extends to a holomorphic map $\phi : \Sigma \to \mathbf{S}^2$. A consequence of Fischer-Colbrie's work is that, in this case, the index of M, as a minimal surface, coincides with the index of its extended Gauss map ϕ, the invariant that we have defined in (1-3) and (1-10). Also, it can be shown, by using elliptic regularity, that the space of the bounded Jacobi fields on the surface M is actually the nullity space $N(\phi)$ of ϕ defined in (1-4). So, as a consequence we have got:

CONCLUSION 1. *If $\phi : \Sigma \to \mathbf{S}^2$ is a holomorphic map from a compact Riemann surface Σ to the two-sphere $\mathbf{S}^2$, the index of ϕ defined in* (1-3) *can be thought of as the index of any complete minimal surface in $\mathbf{R}^3$ with finite total curvature which has ϕ as extended Gauss map, and the nullity space $N(\phi)$ of ϕ is the space consisting of all bounded Jacobi fields on any of these surfaces.*

Willmore surfaces. We will consider now a second geometrical context in which the Schrödinger operator (1-1) associated to a holomorphic map defined on a compact Riemann surface appears again. This is the study of Willmore surfaces, that is, those surfaces immersed into the three-dimensional Euclidean space or into the three-sphere which are critical points of the Willmore functional.

In fact, take a branched conformal immersion $F : \Sigma \to \mathbf{S}^3$ from a compact Riemann surface Σ into the three-sphere. Denote by N a unit normal field for F and by H the mean curvature of F with respect to this N. Then, one can consider *the conformal Gauss map of F* that we will represent by ψ and which is given by:

$$\psi = (H, HF + N) : \Sigma \longrightarrow \mathbf{R} \times \mathbf{R}^4 = \mathbf{R}^5.$$

If we endow the space $\mathbf{R}^5$ with the Minkowski metric $\langle , \rangle$ with signature $(-,+,+,+,+)$ and consider on $\mathbf{C}^5$ its $\mathbf{C}$-bilinear extension, that we will also denote by $\langle , \rangle$, one can get by taking into account that $\langle F, F\rangle = 1$ and $\langle F_z, F_z\rangle = 0$, for a local complex coordinate z on Σ, that

$$\text{(2-1)} \qquad \langle \psi, \psi\rangle = 1 \quad \langle \psi_z, \psi_z\rangle = 0 \quad |\psi_z|^2 = (H^2 - G)|F_z|^2,$$

where G is the determinant of the second fundamental form of F. So, if $\mathbf{S}^4_1$ stands for the set of length $+1$ vectors in the Minkowski space $\mathbf{R}^5$, that is, for the four-dimensional De Sitter space, we have that the conformal Gauss map ψ is actually a weakly conformal map

$$\psi : \Sigma \longrightarrow \mathbf{S}^4_1$$

which is regular away from the umbilical and branching points of F and whose induced area coincides, up to a constant, with the value that the Willmore functional takes on F. This conformal Gauss map had been considered by Blaschke [**Bl**] and, recently, has been rediscovered by Bryant [**Br1**] and its importance lies on the following fact, proved by him in [**Br1**], that one can easily get from (2-1):

$$\text{(2-2)} \qquad \psi_{z\bar{z}} + |\psi_z|^2\psi = 0 \iff H_{z\bar{z}} + |F_z|^2(H^2 - G)H = 0,$$

that is,

$$\psi \text{ is harmonic} \iff F \text{ is a Willmore immersion.}$$

Suppose now that, in fact, F is a critical point of the Willmore functional. Then we can dispose of a map $\psi : \Sigma \to \mathbf{S}_1^4$ which is conformal and harmonic. Remember that, in this type of situation, when the target manifold was an Euclidean four–sphere, a quartic holomorphic form had been constructed by Calabi and Chern (see [**Che**], for instance). So, a good trick to do here is to forget the 1 in $\mathbf{S}_1^4$ and define a quartic form q_F like in the Euclidean case, as follows:

$$q_F = \langle \psi_{zz}, \psi_{zz} \rangle \, (dz)^4$$

where z means a local complex coordinate on the surface Σ. From (2-2), we can see that q_F is holomorphic provided that F is a Willmore immersion (see [**Br1**]). But now there is a difference with the Euclidean case because this quartic form q_F vanishing has quite a strong consequence. One can easily prove that q_F is identically zero if and only if either F is umbilical or the image of the conformal Gauss map ψ lies in a degenerate hyperplane of the Minkowski space, that is

$$q_F = 0 \Longleftrightarrow \begin{cases} F \text{ is umbilical} \\ \exists A \in \mathbf{R}^5 - \{0\}, \ \langle A, A \rangle \, , \ \langle \psi, A \rangle = 0. \end{cases}$$

In this case, choose any Lorentz plane in the Minkowski space containing the vector A and call $\mathbf{R}^3$ to its orthogonal complement which is Euclidean. So, we can write ψ in the following way:

$$\psi = uA + \phi, \quad u \in C^\infty(\Sigma), \quad \phi : \Sigma \to \mathbf{R}^3.$$

But now, we have

$$\begin{aligned} \langle \psi, \psi \rangle = 1 &\Longrightarrow \langle \phi, \phi \rangle = 1 \\ \langle \psi_z, \psi_z \rangle = 0 &\Longrightarrow \langle \phi_z, \phi_z \rangle = 0 \\ \psi_{z\bar{z}} + |\psi_z|^2 \psi = 0 &\Longrightarrow u_{z\bar{z}} + |\phi_z|^2 u = 0. \end{aligned}$$

The first two equations tell us that ϕ maps Σ onto the two–sphere $\mathbf{S}^2$ and that ϕ is holomorphic. The latter can be written in the following way:

$$\Delta_0 u + |\nabla_0 \phi|^2 u = 0$$

where Δ_0 and ∇_0 are respectively the Laplacian and the gradient of the local Euclidean metric $|dz|^2$. Hence, the function u lies in the nullity space $N(\phi)$ of the holomorphic map ϕ defined in (1-4). Furthermore, the reader can see, as a little exercise, that, if $u \in L(\phi)$, see (1-6), then F would be umbilical. Of course, all this process can be inverted and, so, we get the following

CONCLUSION 2. *Each non–umbilical Willmore immersion with Bryant's quartic form identically zero (for example, each spherical Willmore immersion) can be represented by a pair (ϕ, u) consisting of a holomorphic map ϕ to the two–sphere and a non–linear function u lying in the nullity space $N(\phi)$ of ϕ.*

As a consequence, it seems to be interesting in order, for instance, to finish the study of the moduli space of spherical Willmore surfaces, started by Bryant (see [**Br1**] and [**Br2**]), to determine which holomorphic maps from a compact Riemann surface to the two–sphere have nullity greater than three.

Determinant of the Laplacian. There is a third geometrical problem involving the spectral invariants that we have introduced in the first section for a holomorphic map from a compact Riemann surface to the two-sphere. It is the study of the determinant of the Laplacian of metrics on compact surfaces.

This topic was initiated by Polyakov and continued, among others, by Onofri and Virasoro [**On–V**] from a physical point of view, in connection with quantum string theory. In fact, in this theory, one deals with the space of metrics in a conformal class on a given compact surface and the considered action can be described as follows. Take a metric ds^2 on a compact Riemann surface Σ compatible with its complex structure and consider the spectrum of its Laplacian

$$\text{Spec}\,\Delta = \{\lambda_0 = 0 < \lambda_1 < \cdots < \lambda_k < \cdots\}.$$

So, we can construct a functional on this space by mapping the metric ds^2 on the product of all its non-zero eigenvalues

$$ds^2 \longrightarrow \prod_{k=1}^{\infty} \lambda_k,$$

of course, after making sense out of this expression. This can be done, for instance, by using some elementary complex analysis as one can see in [**La**] or in [**Os–Ph–Sa**]. Onofri and Virasoro showed that a metric ds^2 is a solution for the Euler–Lagrange equation of the variational problem corresponding to this functional, when its area is constrained to take a fixed value, if and only if it has constant Gauss curvature.

In this formulation of the Polyakov quantum string theory, due to Onofri and Virasoro, the Gauss–Bonnet theorem imposes a limitation from the physical point of view: if we fix a constant to be the Gauss curvature of a metric on the surface Σ, then its area can take only one possible value. This limitation could be removed by considering metrics with singularities, such as the metric ds^2_ϕ induced on Σ by a holomorphic map ϕ to the two-sphere. All these metrics have constant one Gauss curvature but their areas depend on the degree of the map ϕ. So, this degree could be introduced as a new quantum number in the theory.

Then, if $\phi : \Sigma \to \mathbf{S}^2$ is a holomorphic map from a compact Riemann surface to the two-sphere, we will consider the class C_ϕ of all the compatible metrics on Σ which have the same area and singularities as the metric ds^2_ϕ. By following, for example, [**Os–Ph–Sa**], one can make a study of the determinant of the Laplacian functional on this set C_ϕ and get:

CONCLUSION 3. *The metric ds^2_ϕ induced on a compact Riemann surface Σ by means of a holomorphic map $\phi : \Sigma \to \mathbf{S}^2$ is a critical point of the determinant of the Laplacian functional on the class C_ϕ. Moreover, the Hessian of this functional at the point ds^2_ϕ is nothing but the quadratic form Q_ϕ defined in* (1-2), *restricted to the space of the functions which are orthogonal to the constants. So, the index of the metric ds^2_ϕ as a critical point of the determinant of the Laplacian is* Ind $\phi - 1$.

We think that the interest of the three geometrical situations that we have just refered to is a sufficient motivation to study the index and the nullity of meromorphic functions defined on compact Riemann surfaces. We want now to state some results that we have obtained on this subject.

Nullity space and minimal surfaces with planar ends. We will begin by giving a certain representation for the functions which lie in the nullity space $N(\phi)$ of a holomorphic map ϕ defined on a compact Riemann surface Σ and taking values on the two–sphere. In Conclusion 1 of the second section, we had stated that $N(\phi)$ contains the bounded Jacobi fields on any complete minimal surface M in $\mathbf{R}^3$ with finite total curvature and extended Gauss map ϕ. On the other hand, we know that fields on $\mathbf{R}^3$ whose flow consists of isometries (Killing fields) or dilatations induce on M Jacobi fields. The next result gives conditions on the minimal surface M in order for these Jacobi fields to be bounded. Before ennoncing it, we need some definitions.

Let Σ be a compact Riemann surface and

$$X : M = \Sigma - \{p_1, \dots, p_k\} \longrightarrow \mathbf{R}^3$$

a minimal immersion with finite total curvature. The points $p_1, \dots, p_k$ are called *the ends of* X. Such an end $p_i \in \Sigma$ is *embedded* if X is injective on a punctured neighbourhood of p_i on Σ. It is said to be *planar* if there exists $a_i \in \mathbf{S}^2$ such that the harmonic function $\langle X, a_i \rangle$ is bounded near to p_i and said to be of *catenoid type* if there exists $a_i \in \mathbf{S}^2$ such that $\langle X, a_i \rangle$ has a logarithmic singularity at p_i. Notice that an embedded end is obliged to be planar or of catenoid type (see [**Sch**]) and that, in the literature, authors use these denominations only for embedded ends.

PROPOSITION 1. *Let Σ be a compact Riemann surface and $X : \Sigma - \{p_1, \dots, p_k\} \to \mathbf{R}^3$ a complete minimal immersion with finite total curvature whose extended Gauss map is the holomorphic map $\phi : \Sigma \to \mathbf{S}^2$. We have that*

a) *If all the ends of X are either planar or of catenoid type and there exists $a \in \mathbf{S}^2$ such that $\phi(p_i) = \pm a$, $i = 1, \dots, k$, then $\langle X \wedge \phi, a \rangle \in N(\phi)$.*
b) *If all the ends of X are planar, then $\langle X, \phi \rangle \in N(\phi)$.*

PROOF: As we have pointed out above, we only need to prove that these two functions defined on $\Sigma - \{p_1, \dots, p_k\}$ are bounded near to the ends $p_1, \dots, p_k$ of X. Fix $i \in \{1, \dots, k\}$ and suppose, by rotating $\mathbf{S}^2$, that $\phi(p_i)$ is the north pole. Now, we identify ϕ, via stereographic projection from this north pole, with a meromorphic function $g : \Sigma \to \overline{\mathbf{C}}$. So

$$\phi = \frac{1}{1 + |g|^2} \left(g + \overline{g}, -i(g - \overline{g}), -1 + |g|^2 \right) \tag{3-1}$$

and, by using the Weierstrass representation, we have

$$X = \frac{1}{2} \Re \int \left(1 - g^2, i(1 + g^2), 2g \right) \omega \tag{3-2}$$

where ω is a meromorphic differential on Σ which is holomorphic away from the ends $p_1, \ldots, p_k$. As $g(p_i) = \infty$, we can choose a local complex coordinate z on a certain neighbourhood D_i of p_i in Σ such that $z(p_i) = 0$ and

$$g(z) = \frac{1}{z^m}, \quad m \geq 1, \quad z \in D_i.$$

With respect to this local coordinate, we have

$$\omega = \left(\frac{a}{z^r} + \frac{b}{z^{r-1}} + \cdots + \frac{c}{z} + h(z)\right) dz, \ z \in D_i, \ r \in \mathbf{Z}$$

where $a \in \mathbf{C} - \{0\}$ and h is holomorphic on D_i. Hence, as $r + 2m \geq 2$ because X is complete, if $z \in D_i$

$$X(z) = (1 - r - 2m)\left(\Re \frac{a}{z^{r+2m-1}}, \Im \frac{a}{z^{r+2m-1}}, 0\right) + Y(z) \tag{3-3}$$

where $Y(z)$ contains terms of lower growth.

On the other hand, we know that there exists $a_i \in \mathbf{S}^2$ such that $\langle X, a_i\rangle$ is either bounded or has a logarithmic singularity at p_i. In both two cases, as $r + 2m - 1 \geq 1$ and $a \neq 0$, we have from (3-3) that $a_i = (0, 0, \pm 1)$, that is, the third coordinate of X is bounded in the case b) or has a logarithmic singularity in the case a). We will deal with each case separately.

Case a). We have that

$$\langle X \wedge \phi, a_i\rangle = \det\big(X, \phi, \pm(0,0,1)\big) = \pm \det(\tilde{X}, \tilde{\phi})$$

where $\tilde{X}$ and $\tilde{\phi}$ are the orthogonal projections of X and ϕ on the plane $x_3 = 0$ in $\mathbf{R}^3$. So, from (3-1) and (3-3), we get

$$\langle X \wedge \phi, a_i\rangle = (1 - r - 2m)\Re \frac{ia}{z^{r+m-1}} + W(z) \tag{3-4}$$

where W includes only terms of lower growth. As the third coordinate of X has at p_i a logarithmic singularity and

$$X_3 = \Re \int \omega g = \Re \int \frac{1}{z^m}\left(\frac{a}{z^r} + \ldots\right) dz,$$

then $m + r \leq 1$, and using (3-4) we conclude.

Case b). In this case the third coordinate of X is bounded at p_i and, so, $m + r \leq 0$. On the other hand, using (3-1) and (3-3), one has

$$\left\langle \tilde{X}, \tilde{\phi}\right\rangle = (1 - r - 2m)\Re \frac{a}{z^{r+m-1}} + W'(z)$$

and $X_3\phi_3$ is bounded from our hypothesis. So $\langle X, \phi\rangle = \left\langle \tilde{X}, \tilde{\phi}\right\rangle + X_3\phi_3$ is bounded.

REMARK. The result above is also true for complete *branched* minimal surfaces. Moreover, notice that the hypothesis in a) is automatically satisfied when X is an embedding (see [J–M] for example).

Now, let $\phi : \Sigma \to \mathbf{S}^2$ be a non–constant holomorphic map. In terms of a local complex coordinate z, the fact that ϕ is holomorphic is given by

$$\langle \phi_z, \phi_z \rangle = 0$$

where $\langle , \rangle$ is the usual bilinear complex product on $\mathbf{C}^3$. By using this equality and $\langle \phi, \phi \rangle = 1$, we obtain the following equations:

$$\text{(3-5)} \qquad \begin{array}{ll} \langle \phi, \phi_z \rangle = 0 & \phi_{zz} = (\partial \log |\phi_z|^2)\phi_z \\ \phi_{z\bar{z}} + |\phi_z|^2 \phi = 0 & \partial\bar{\partial} \log |\phi_z|^2 = -|\phi_z|^2 \end{array}$$

where ∂ and $\bar{\partial}$ denote differentiation with respect to the variables z and $\bar{z}$ respectively. Take a function $u \in N(\phi)$ in the nullity space of ϕ defined in (1-4). By using on Σ the Euclidean local metric $|dz|^2$ we have that

$$u_{z\bar{z}} + |\phi_z|^2 u = 0.$$

Let $B(\phi) = m_1 p_1 + \cdots + m_k p_k$ be the ramification divisor of ϕ. Then we define a map

$$X(u) : \Sigma - \{p_1, \ldots, p_k\} \longrightarrow \mathbf{R}^3$$

by

$$\text{(3-6)} \qquad X(u) = u\phi + \frac{1}{|\phi_z|^2} \{u_z \phi_{\bar{z}} + u_{\bar{z}} \phi_z\}.$$

Clearly $X(u)$ is a real vector and the expression on the right side of (3-6) is independent of the choice of the isothermal parameter z. Differentiating with respect to z, one obtains, by using (3-5)

$$\text{(3-7)} \qquad X(u)_z = \left(u_{zz} - (\partial \log |\phi_z|^2) u_z\right) \frac{\phi_{\bar{z}}}{|\phi_z|^2}.$$

Notice that $X(u)$ is a constant vector if $u \in L(\phi)$ defined in (1-6). Hence

$$\langle X(u)_z, \phi \rangle = 0 \quad \langle X(u)_z, \phi_{\bar{z}} \rangle = 0 \quad \langle X(u)_z, X(u)_z \rangle = 0.$$

In particular, $X(u)$ is a conformal map from $\Sigma - \{p_1, \ldots, p_k\}$ to $\mathbf{R}^3$. Moreover, by using (3-5) again

$$\begin{aligned} 0 &= \langle X(u)_z, \phi \rangle_{\bar{z}} = \langle X(u)_{z\bar{z}}, \phi \rangle \\ 0 &= \langle X(u)_z, \phi_{\bar{z}} \rangle_{\bar{z}} = \langle X(u)_{z\bar{z}}, \phi_{\bar{z}} \rangle . \end{aligned}$$

Since $X(u)_{z\bar{z}}$ is real, we also have that $\langle X(u)_{z\bar{z}}, \phi_z\rangle = 0$ and, so

$$X(u)_{z\bar{z}} = 0,$$

that is, $X(u)$ is harmonic.

As a conclusion, the map $X(u) : \Sigma - \{p_1, \ldots, p_k\} \to \mathbf{R}^3$ is a branched minimal immersion whose Gauss map is ϕ and whose support function is $\langle X(u), \phi\rangle = u$. To study the behaviour of the minimal immersion $X(u)$ at its ends $p_1, \ldots, p_k \in \Sigma$, it is convenient to identify ϕ with a meromorphic function $g : \Sigma \to \overline{\mathbf{C}}$ as in (3-1) and to use the Weierstrass representation (3-2) for $X(u)$. So, we have

$$X(u)_z\, dz = \frac{\omega}{2}\left(1 - g^2, i(1 + g^2), 2g\right).$$

Comparing this equation with (3-7), we can see that the meromorphic differential ω is

$$\omega = \frac{1}{g'}\left(u_{zz} + \left(\frac{2\bar{g}g'}{1 + |g|^2} - \frac{g''}{g'}\right)u_z\right)\, dz.$$

It is easy to prove that, at the point p_i, $i = 1, \ldots, k$, there exists a possible pole of ω with order less than or equal to $m_i + 1$. Then, the meromorphic quadratic differential σ on Σ given by

$$\sigma = \omega\, dg \tag{3-8}$$

has a possible pole with order less than or equal to one at each end p_i. Let z be a local complex coordinate on a certain neighbourhood of the point p_i on Σ such that $z(p_i) = 0$ and suppose that on this neighbourhood we have, up to a rotation of $\mathbf{R}^3$

$$g(z) = z^{m_i+1}, \quad m_i \geq 1.$$

So, locally, from (3-8)

$$X(u)_z\, dz = \frac{\sigma}{2\, dg}\left(1 - g^2, i(1 + g^2), 2g\right) =$$

$$\frac{1}{2(m_i + 1)}\left(\frac{1}{z^{m_i}} - z^{m_i+2}, i\left(\frac{1}{z^{m_i}} + z^{m_i+2}\right), 2z\right)\left(\frac{a}{z} + \sum_{i\geq 0} a_i z^i\right) dz$$

where

$$\left(\frac{a}{z} + \sum_{i\geq 0} a_i z^i\right)(dz)^2$$

is a local expression for σ around the point p_i. Then, the third coordinate of $X(u)_z\, dz$ is holomorphic at p_i and, so, $\langle X, (0,0,1)\rangle$ is bounded near to p_i. Moreover, as

$$X(u) = \Re \int X(u)_z\, dz,$$

we have $\Im\ \mathrm{Res}_{p_i} X(u)_z\, dz = 0$, that is $\Im a_{m_i-1}(1, i, 0) = 0$. So, $a_{m_i-1} = 0$ and

$$\mathrm{Res}_{p_i} X(u)_z\, dz = 0.$$

As a consequence:

Proposition 2. *Let Σ be a compact Riemann surface and $\phi : \Sigma \to \mathbf{S}^2$ a holomorphic map. If $u \in N(\phi) - L(\phi)$, then the map $X(u)$ defined in (3-6) is a complete branched minimal immersion with finite total curvature and planar ends whose extended Gauss map is ϕ and whose support function is u.*

If we denote by $M(\phi)$ the linear space of all the complete branched minimal immersions (including the constant maps) into $\mathbf{R}^3$ with finite total curvature and planar ends whose extended Gauss map is ϕ, we have by taking into account the two linear maps:

$$\begin{aligned} u \in N(\phi) &\longmapsto X(u) \in M(\phi), \text{ see (3-6)}, \\ X \in M(\phi) &\longmapsto \langle X, \phi \rangle \in N(\phi), \end{aligned} \tag{3-9}$$

the following result:

Theorem 3. *Let Σ be a compact Riemann surface and $\phi : \Sigma \to \mathbf{S}^2$ a holomorphic map. Then, the linear spaces $N(\phi)$ and $M(\phi)$ are isomorphic via the linear maps given in (3-9). Moreover, we have an induced isomorphism*

$$\frac{N(\phi)}{L(\phi)} \cong \frac{M(\phi)}{\text{constants}}.$$

Proof: It suffices to prove that the linear maps given in (3-9) are inverse to each other and use Proposition 1 and Proposition 2. The second assertion is nothing but a consequence of the fact that the first map in (3-9) applies the linear functions of $L(\phi)$ defined in (1-6) onto the subspace of $M(\phi)$ consisting of the constants.

Remark. Combining Conclusion 2 and this Theorem 3, one can obtain a relation between branched Willmore surfaces in $\mathbf{S}^3$ and complete branched minimal surfaces in $\mathbf{R}^3$ with finite total curvature and planar ends. This relation was obtained by Bryant in [**Br1**] in the non–ramified case.

By using the Weierstrass representation (3-2), we can identify each element X of $M(\phi)$ with a pair (g, ω), where g is the meromorphic function $g : \Sigma \to \overline{\mathbf{C}}$ associated to ϕ by means of (3-1) and ω is a meromorphic differential on Σ which is holomorphic on $\Sigma - \{p_1, \ldots, p_k\}$, where $B(\phi) = m_1 p_1 + \cdots + m_k p_k$ is the ramification divisor of ϕ. If $q_1, \ldots, q_d \in \Sigma$ are the poles of g, we define a divisor $D(\phi)$ on Σ by

$$D(\phi) = \sum_{i=1}^{k} (m_i + 1) p_i - 2 \sum_{i=1}^{d} q_i.$$

Then, Theorem 3 above can be reformulated in the following way:

Theorem 4. *The linear space $N(\phi)/L(\phi)$ is isomorphic to*

$$\left\{ \omega \in H^0(\kappa_\Sigma \otimes [D(\phi)]) \,|\, \mathrm{Res}_{p_i} \omega = 0, \ \Re \int_\gamma \omega \left(1 - g^2, i(1 + g^2), 2g\right) = 0, \ \gamma \in H_1(\Sigma, \mathbf{Z}) \right\}$$

where κ_Σ is the canonical line bundle on Σ and $[D(\phi)]$ is the line bundle associated to the divisor $D(\phi)$.

Notice that the space $N(\phi)/L(\phi)$ can also be described in terms of the meromorphic quadratic differential σ defined in (3-8). In fact

THEOREM 5. *The linear space $N(\phi)/L(\phi)$ is isomorphic to*

$$\left\{\sigma \in H^0\left(2\kappa_\Sigma \otimes \left[\sum_{i=1}^{k} p_i\right]\right) \mid \left\{ \begin{array}{l} Res_{p_i} \dfrac{\sigma}{dg} = 0, \\ \Re \displaystyle\int_\gamma \dfrac{\sigma}{dg}\,(1-g^2, i(1+g^2), 2g) = 0,\ \gamma \in H_1(\Sigma, \mathbf{Z}) \end{array} \right. \right\}.$$

Holomorphic maps of low index. We shall begin by gathering some information concerning the index of the holomorphic maps from a compact Riemann surface to the two–sphere and by trying to answer the following natural question: What can one say about these holomorphic maps which have the least index? Of course, we are excluding the trivial case of index zero. So, we are looking for the properties that such a holomorphic map must satisfy in order to have index one. This case is doubly interesting because it includes, from Conclusions 1 and 3, complete minimal surfaces of finite total curvature with the lowest index and branched metrics which are local maxima of the detrminant of the Laplacian functional. These metrics should provide classical models of vacuum states for the Polyakov quantum string theory.

THEOREM 6. *Let Σ be a compact Riemann surface and $\phi : \Sigma \to \mathbf{S}^2$ a holomorphic map with index one. Then, there are no holomorphic maps from Σ to $\mathbf{S}^2$ with degree less than $\deg \phi$. Moreover, if $\psi : \Sigma \to \mathbf{S}^2$ is a holomorphic map with $\deg \psi = \deg \phi$, then $\psi = G \circ \phi$ for a certain conformal transformation G of $\mathbf{S}^2$.*

PROOF: From the definition (1-2), we have

$$Q_\phi(1,1) = -\int_\Sigma |\nabla\phi|^2\, dA = -8\pi \deg \phi < 0,$$

since ϕ cannot be a constant map. As the index of ϕ is one, we obtain, from (1-3), that the form Q_ϕ is positive semidefinite on the hyperplane of $W_1(\Sigma)$ orthogonal to the constants with respect to the bilinear form associated to Q_ϕ. That is,

$$\int_\Sigma \{|\nabla u|^2 - |\nabla\phi|^2 u^2\}\, dA \geq 0 \quad \text{if} \quad \int_\Sigma |\nabla\phi|^2 u\, dA = 0$$

and the equality happens if and only if $u \in N(\phi)$.

Let $\psi : \Sigma \to \mathbf{S}^2$ be any holomorphic map. As $|\nabla\phi|^2$ is positive except at a finite set, we can use a known result by Hersch **[He]** to obtain a conformal transformation C of $\mathbf{S}^2$ such that

$$\int_\Sigma |\nabla\phi|^2 (C \circ \psi)\, dA = 0.$$

Then

$$0 \leq \int_\Sigma \{|\nabla(C\circ\psi)|^2 - |\nabla\phi|^2 |C\circ\psi|^2\}\, dA =$$
$$\int_\Sigma \{|\nabla\psi|^2 - |\nabla\phi|^2\}\, dA = 8\pi(\deg\psi - \deg\phi).$$

Hence $\deg\phi \leq \deg\psi$ for each $\psi : \Sigma \to \mathbf{S}^2$ holomorphic. If the equality held, then we would have that $\langle C\circ\psi, a\rangle \in N(\phi)$ for all $a \in \mathbf{R}^3$ and, as $\langle C\circ\psi, a\rangle \in N(C\circ\psi)$, we would obtain that $|\nabla\phi|^2 = |\nabla(C\circ\psi)|^2$. By using Lemma 7 below one would get an isometry A of $\mathbf{S}^2$ such that $\phi = A \circ C \circ \psi$.

LEMMA 7. *Let Σ be a compact Riemann surface and $\phi, \psi : \Sigma \to \mathbf{S}^2$ two holomorphic maps such that $|\nabla\phi|^2 = |\nabla\psi|^2$ with respect to any metric compatible with the complex structure. Then, there exists an isometry $A : \mathbf{S}^2 \to \mathbf{S}^2$ such that $\psi = A \circ \phi$.*

PROOF: As $|\nabla\phi|^2 = |\nabla\psi|^2$, we have $N(\phi) = N(\psi)$ from the definition (1-4). Fix $a \in \mathbf{S}^2$ and put $u = \langle \psi, a \rangle$. Then $u \in N(\phi)$ from (1-6). Hence, the map $X(u)$ defined in (3-6) is harmonic on $\Sigma - \{p_1, \ldots, p_k\}$, where $p_1, \ldots, p_k$ are the ramification points of ϕ. But, in our case,

$$\begin{aligned} |X| &= u^2 + 2\frac{|u_z|^2}{|\phi_z|^2} = \langle \psi, a \rangle^2 + 2\frac{|\langle \psi_z, a \rangle|^2}{|\phi_z|^2} \\ &\leq 1 + 2\frac{|\psi_z|^2}{|\phi_z|^2} = 3. \end{aligned}$$

So, X is bounded and, hence, is a constant vector map $b \in \mathbf{R}^3$. Then

$$\langle \psi, a \rangle = u = \langle X, \phi \rangle = \langle b, \phi \rangle .$$

This occurs for each $a \in \mathbf{S}^2$. Taking into account that $|\psi|^2 = |\phi|^2 = 1$, one can conclude the proof.

From Theorem 6, we are led to consider another question: Which is the least degree that a meromorphic function can have on a given compact Riemann surface? And, how many meromorphic functions, up to a Möbius transformation, attain that degree? This is an old problem ant the Brill–Noether theorem tries to answer it (see [**Fa–Kr**] or [**Gr–Ha**]). In fact, this theorem asserts that, for the generic compact Riemann surface of genus greater than two, the least possible degree is

$$1 + \left[\frac{1 + \text{genus } \Sigma}{2}\right]$$

and that there is no uniqueness up to Möbius transformations. The cases of genus zero or one are trivial and, so, we have the following result.

COROLLARY 8. *Let $\phi : \Sigma \to \mathbf{S}^2$ a holomorphic map defined on a compact Riemann surface. We have:*

a) *If* genus $\Sigma = 0$, *then* Ind $\phi = 1 \Leftrightarrow \deg \phi = 1$.
b) *If* genus $\Sigma = 1$, *then* Ind $\phi = 1$ *is impossible.*
c) *For the generic complex structure with genus greater than two,* Ind $\phi = 1$ *is impossible.*
d) *In all the cases* $\deg \phi \leq 1 + \left[\frac{1+\text{genus } \Sigma}{2}\right]$, *provided that Ind* $\phi = 1$.

Suppose now that the holomorphic map $\phi : \Sigma \to \mathbf{S}^2$ that we are considering is the extended Gauss map of a complete minimal surface in $\mathbf{R}^3$ with finite total curvature. Then, we have a non–trivial relation between the degree of ϕ and the genus of the surface Σ. In fact, we know from a result by Osserman [**Oss**] that

$$\deg \phi \geq 1 + \text{genus } \Sigma.$$

So, combining this information with d) in Corollary 8, one has two possible genera for the surface Σ, say zero or one. This latter cannot occur from b). Then, Σ must have genus zero and the map ϕ degree one. Now, we can conclude from another result by Osserman [**Oss**] the following statement.

Corollary 9. [**Lo–R**]. *The catenoid and the Enneper surface are the only complete minimal surfaces in* $\mathbb{R}^3$ *with index one.*

With respect to the determinant of the Laplacian functional, we must remark that we do not know any holomorphic map, in the non–zero genus case, with index one. But, according to Theorem 7, one should look for them, for instance, on a hyperelliptic Riemann surface. In fact, we think that the two–covering map from such a surface to the two–sphere has index one, provided that the images of its Weierstrass points are spread enough on the sphere.

Estimates for the index and the nullity. The case of low index above points out that there is a strong connection between the index of a holomorphic map'and its degree. In fact, in a final remark of her work about index of minimal surfaces in three–dimensional manifolds [**FC**], Fischer-Colbrie said that one should be able to get an explicit relation between the index of a holomorphic map from a compact Riemann surface to the two–sphere and its geometry. Even more, Gulliver [**Gu**] suggested a concrete inequality involving the index and the degree of such a holomorphic map and the genus of the surface. In the last five years, a lot of work have been dedicated to obtain this explicit relation. For arbitrary genus, the first approach was done by Tysk in [**Ty**]. He obtained the following estimate:

Theorem 10. [**Ty**]. *If* $\phi : \Sigma \to \mathbf{S}^2$ *is a holomorphic map defined on a compact Riemann surface, then*

$$\operatorname{Ind} \phi < 7.7 \deg \phi.$$

The proof is based on comparing the heat kernel of the standard metric ds_0^2 on $\mathbf{S}^2$ and that of the induced metric ds_ϕ^2 on Σ, see (1-7). This is probably a rough estimate as we will see in the sequel.

Notice that any assertion about the index and the nullity os a holomorphic map $\phi : \Sigma \to \mathbf{S}^2$ can be read as an assertion about the multiplicities of some eigenvalues of the Laplacian Δ_ϕ of the branched metric ds_ϕ^2, as we had remarked in (1-10) and (1-11). From this point of view and by using standard methods such as Courant's nodal theorem [**Co–Hi**], we will obtain some estimates for the index and the nullity of the holomorphic map ϕ when its branching values are in especial positions on the sphere. In the particular case in which all the branching values lie in an equator, we will get a sharp result about the multiplicities of the eigenvalues of Δ_ϕ which come from the spectrum of the standard two–sphere. As a consequence, we will compute the index and the nullity for such a holomorphic map.

We will start with the following more or less well–known result:

Lemma 11. *Let* Ω *be a two–dimensional half–sphere of radius one endowed with the standard metric* ds_0^2. *Then, we have that:*

a) *The eigenvalues of the Laplacian for the Neumann problem on* Ω *are* $\lambda_k = k(k+1)$ *with multiplicities* $m_k = k+1$, $k = 0, 1, 2, \ldots$

b) *The eigenvalues of the Laplacian for the Dirichlet problem on* Ω *are* $\mu_k = k(k+1)$ *with multiplicities* $n_k = k$, $k = 1, 2, 3 \ldots$

SKETCH OF PROOF: Denote by $A : \mathbf{S}^2 \to \mathbf{S}^2$ the symmetry with respect to the plane where the equator of Ω lies. Then, each eigenfunction for the Neumann problem on Ω can be extended on the whole of $\mathbf{S}^2$ to obtain an A-invariant eigenfunction of the Laplacian of ds_0^2. Also, each eigenfunction for the Dirichlet problem on Ω provides us, after a suitable extension, an A-antiinvariant eigenfunction of the Laplacian of ds_0^2. Now, it suffices to know that $k(k+1)$, $k = 0, 1, 2, \ldots$, are the eigenvalues of the standard metric ds_0^2 on $\mathbf{S}^2$ with multiplicities $2k+1$ and to analyse which of the corresponding eigenfunctions are either A-invariant or A-antiinvariant.

Now, if $\phi : \Sigma \to \mathbf{S}^2$ is a holomorphic map defined on a compact Riemann surface, from the definition (1-7) of the metric ds_ϕ^2 induced on Σ by ϕ, we have that

$$\operatorname{Spec} \Delta_0 \subset \operatorname{Spec} \Delta_\phi$$

where Δ_0 is the Laplacian of the standard metric ds_0^2 on $\mathbf{S}^2$. Then, for $k = 0, 1, 2, \ldots$

$$k(k+1) \in \operatorname{Spec} \Delta_\phi$$

with multiplicity at least $2k+1$. The following computations seek to staudy which is the exact multiplicity of $k(k+1)$ as an eigenvalue of Δ_ϕ and how many eigenvalues of Δ_ϕ are between $k(k+1)$ and $(k+1)(k+2)$.

LEMMA 12. *Let $\Omega_1, \ldots, \Omega_m$ be a partition of a compact Riemann surface Σ by domains with piecewise smooth boundary. If $\phi : \Sigma \to \mathbf{S}^2$ is a non-constant holomorphic map, then*

$$\#\{\Delta_\phi\text{-eigenvalues } \lambda \text{ on } \Sigma \text{ with } \lambda < k(k+1)\} \geq$$

$$\sum_{i=1}^{m-1} \#\{\Delta_\phi\text{-eigenvalues } \lambda \text{ for the Dirichlet problem on } \Omega_i \text{ with } \lambda \leq k(k+1)\} +$$

$$\#\{\Delta_\phi\text{-eigenvalues } \lambda \text{ for the Dirichlet problem on } \Omega_m \text{ with } \lambda < k(k+1)\}$$

for any $k = 1, 2, 3, \ldots$

PROOF: Fix $k = 1, 2, 3, \ldots$ and denote by V_i, $i = 1, \ldots, m-1$, the linear space spanned by the eigenfunctions of Δ_ϕ for the Dirichlet problem on the domain Ω_i whose corresponding eigenvalues are less than or equal to $k(k+1)$. In the same way, let V_m be the linear space spanned by the eigenfunctions of Δ_ϕ for the Dirichlet problem on Ω_m with eigenvalues less than $k(k+1)$. These $V_1, \ldots, V_{m-1}, V_m$ can be viewed as subspaces of the Sobolev space $W_1(\Sigma)$ by extending each function that they contain by zero on the whole of Σ. Finally, we denote by V the linear subspace of $W_1(\Sigma)$ spanned by the eigenfunctions of Δ_ϕ on Σ with corresponding eigenvalues less than $k(k+1)$. Afetr this, the assertion that we want to prove in this lemma can be written in the following way:

$$\dim V \geq \sum_{i=1}^{m} \dim V_i. \tag{3-10}$$

But, as a direct consequence of the definitions of these linear spaces, we have that, for any function $u \in W_1(\Sigma)$ which is $L^2(\Sigma)$-orthogonal to V with respect to the metric ds^2_ϕ

$$\int_\Sigma |\nabla u|^2 \, dA_\phi \geq k(k+1) \int_\Sigma u^2 \, dA_\phi \tag{3-11}$$

and the equality holds if and only if u is a $k(k+1)$-eigenfunction of Δ_ϕ on Σ. Also, for any $u \in \bigoplus_{i=1}^m V_i$, we have

$$k(k+1) \int_\Sigma u^2 \, dA\phi \leq \int_\Sigma |\nabla u|^2 \, dA_\phi \tag{3-12}$$

and the equality holds if and only if the V_i-component of u is a $k(k+1)$-eigenfunction of Δ_ϕ for the Dirichlet problem on Ω_i, $i = 1, \ldots, m$. In particular, the V_m-component of u vanishes.

If the inequality (3-10) were not true, then we could find a non–zero function $u \in \bigoplus_{i=1}^m V_i$ and $L^2(\Sigma)$-ortyhogonal to V. This function would satisfy (3-11) and (3-12). Hence, it would be a $k(k+1)$-eigenfunction of Δ_ϕ on Σ identically zero on Ω_m. By using the unique continuation principle [A], u would vanish on Σ. This contradiction proves (3-10).

LEMMA 13. *Under the same hypothesis as Lemma 12, we have that*

$$\#\{\Delta_\phi\text{-eigenvalues } \lambda \text{ on } \Sigma \text{ with } \lambda \leq k(k+1)\} \leq$$

$$\sum_{i=1}^{m-1} \#\{\Delta_\phi\text{-eigenvalues } \mu \text{ for the Neumann problem on } \Omega_i \text{ with } \mu < k(k+1)\} +$$

$$\#\{\Delta_\phi\text{-eigenvalues } \mu \text{ for the Neumann problem on } \Omega_m \text{ with } \mu \leq k(k+1)\}$$

for any $k = 0, 1, 2, \ldots$

PROOF: For a fixed $k = 0, 1, 2, \ldots$, let V_i be the linear space spanned by the eigenfunctions of Δ_ϕ for the Neumann problem on Ω_i, $i = 1, \ldots, m-1$, with eigenvalues less than $k(k+1)$ and V_m be the space spanned by the eigenfunctions of Δ_ϕ for the Neumann problem on Ω_m with corresponding eigenvalues less than or equal to $k(k+1)$. Then, if $u \in W_1(\Sigma)$ is $L^2(\Sigma)$-orthogonal to each V_i, for $i = 1, \ldots, m$, we have

$$\begin{aligned} &\int_\Sigma |\nabla u|^2 \, dA\phi = \sum_{i=1}^m \int_{\Omega_i} |\nabla u|^2 \, dA\phi \geq \\ &k(k+1) \sum_{i=1}^m \int_{\Omega_i} u^2 \, dA_\phi = k(k+1) \int_\Sigma u^2 \, dA_\phi \end{aligned} \tag{3-13}$$

and the equality holds if and only if $u_{|\Omega_i}$ is a $k(k+1)$-eigenfunction of Δ_ϕ for the Neumann problem on Ω_i, $i = 1, \ldots, m$, and, so, $u_{|\Omega_m} \equiv 0$.

On the other hand, if V denotes the subspace of $W_1(\Sigma)$ spanned by the eigenfunctions of Δ_ϕ on Σ with eigenvalues less than or equal to $k(k+1)$, then each function $u \in V$ satisfies

$$\int_\Sigma |\nabla u|^2 \, dA_\phi \le k(k+1) \int_\Sigma u^2 \, dA_\phi \tag{3-14}$$

and the equality holds if and only if u is a $k(k+1)$-eigenfunction of Δ_ϕ on Σ. But the claim of this lemma can be rewritten in this way

$$\dim V \le \sum_{i=1}^{m} \dim V_i.$$

If this inequality failed to be true, we could choose $u \in V - \{0\}$ and such that $u_{|\Omega_i}$ would be orthogonal to V_i for each $i = 1, \ldots, m$. Hence, the function u would satisfy (3-13) and (3-14) and be a $k(k+1)$-eigenfunction of Δ_ϕ on Σ with $u_{|\Omega_m} \equiv 0$. Then, we prove the lemma by using again the unique continuation principle.

These two lemmae can be combined in order to obtain an accurate estimate in the following theorem, where we will find a partition of the surface Σ by domains which have a good behaviour with respect to the Dirichlet and the Neumann problem corresponding to the operator Δ_ϕ.

THEOREM 14. *Let Σ be a compact Riemann surface and $\phi : \Sigma \to \mathbf{S}^2$ a non–constant holomorphic map. If all the branching values of ϕ lie in an equator of $\mathbf{S}^2$, then, for $k = 0, 1, 2, \ldots$*

a) *The multiplicity of $k(k+1)$ as an eigenvalue of Δ_ϕ on Σ is $2k+1$.*
b) *The number of eigenvalues λ of Δ_ϕ on Σ such that $k(k+1) < \lambda < (k+1)(k+2)$ is $2(k+1)(\deg \phi - 1)$.*

PROOF: We will denote by C the equator of $\mathbf{S}^2$ which contains all the branching values of ϕ and put $d = \deg \phi$. Then $\Sigma - \phi^{-1}(C)$ has exactly $2d$ connected components $\Omega_1, \ldots, \Omega_{2d}$ in such a way that $\phi_{|\Omega_i}$ is an isometry onto a half–sphere for each i. For any $k = 0, 1, 2, \ldots$, it follows from Lemma 11 that

$$\sum_{i=1}^{2d-1} \#\{\Delta_\phi\text{-eigenvalues } \lambda \text{ for the Dirichlet problem on } \Omega_i \text{ with } \lambda \le k(k+1)\} +$$
$$\#\{\Delta_\phi\text{-eigenvalues } \lambda \text{ for the Dirichlet problem on } \Omega_m \text{ with } \lambda < k(k+1)\} =$$
$$k(k+1)d - k,$$

and that

$$\sum_{i=1}^{2d-1} \#\{\Delta_\phi\text{-eigenvalues } \mu \text{ for the Neumann problem on } \Omega_i \text{ with } \mu < k(k+1)\} +$$
$$\#\{\Delta_\phi\text{-eigenvalues } \mu \text{ for the Neumann problem on } \Omega_m \text{ with } \mu \le k(k+1)\} =$$
$$k(k+1)d + k + 1.$$

From these observations, we obtain by using Lemma 12 and Lemma 13 that

$$\#\{\Delta_p hi\text{-eigenvalues } \lambda \text{ on } \Sigma \text{ with } \lambda < k(k+1)\ \} \geq k(k+1)d - k$$
$$\#\{\Delta_\phi\text{-eigenvalues } \lambda \text{ on } \Sigma \text{ with } \lambda \leq k(k+1)\} \leq k(k+1)d + k + 1.$$

As we know that the multiplicity of $k(k+1)$ as an eigenvalues of Δ_ϕ on Σ is at least $2k+1$, we get from the above inequalities that

$$\#\{\Delta_\phi\text{-eigenvalues } \lambda \text{ on } \Sigma \text{ with } \lambda < k(k+1)\ \} = k(k+1)d - k.$$
$$\text{multiplicity of the } \Delta_\phi\text{-eigenvalue } k(k+1) \text{ on } \Sigma = 2k+1.$$

Now the proof of the theorem follows from a simple arithmetic.

As a direct consequence of Theorem 14, we can compute the index and the nullity defined in (1-3) and (1-5) for a holomorphic map such that all its branching values are mapped on an equator of the two–sphere.

COROLLARY 15. *If $\phi : \Sigma \to \mathbf{S}^2$ is a non–constant holomorphic map defined on a compact Riemann surface Σ all of whose branching values lie in an equator of $\mathbf{S}^2$, then*

$$\text{Nul}\ \phi = 3 \qquad \text{Ind}\ \phi = 2\deg\phi - 1.$$

Partial versions of this Corollary have been obtained by Li and Tam, see [**Li–Tr**], Nayatani [**N**] and Choe [**Cho**].

This result has a first obvious application, taking into account Conclusion 1, to compute the index of complete minmal surfaces in $\mathbf{R}^3$ with finite total curvature whose extended Gauss map behaves in that especial way. For instance, the index of the Jorge–Meeks $(k+1)$-catenoid described in [**J–M**] is $2k+1$ because, in this case, $\Sigma = \overline{\mathbf{C}}$ and $\phi(z) = z^k$ and the index of the genus one Chen–Gackstätter surface, recently characterised by López [**Lo**], is three because $\Sigma = \mathbf{C}/\{1, i\}$ and $\phi \equiv a\wp'/\wp$ where $a \in \mathbf{R}$ and $\wp$ is the Weierstrass function associated to the lattice spanned by $\{1, i\}$ in $\mathbf{C}$.

Another application is the following: let $X : M = \Sigma - \{p_1, \ldots, p_k\} \to \mathbf{R}^3$ be a non–flat complete minimal immersion with finite total curvature and suppose that all the branching values of its extended Gauss map ϕ lie in an equator of the sphere. From Corollary 15 and (1-6), we have that the nullity space $N(\phi)$ of ϕ coincides with $L(\phi) = \{\langle\phi, a\rangle \mid a \in \mathbf{R}^3$. If all the ends of X were planar, then Proposition 1 would assert us that the support function $\langle X, \phi\rangle$ of the surface belongs to $L(\phi)$. But this is impossible from Theorem 3. In a similar way, if all the ends of X were parallel and either planar or of catenoid type, then, from Proposition 1, we would have that the function $\langle X \wedge \phi, a\rangle$ lies in $L(\phi)$. But it is a classical result that this occurs if and only if X is a finite Riemannian covering of the catenoid. Hence, we can state:

COROLLARY 16. *Let M be a non–flat complete minimal surface in $\mathbf{R}^3$ with finite total cutvature such that all the branching values of its extended Gauss map lie in an equator of $\mathbf{S}^2$. Then*

a) *M has at least a non–planar end.*
b) *If all the ends of M are parallel and either planar or of catenoid type, then M is a finite Riemannian covering of the catenoid.*

For later applications we need to estimate the index and the nullity for holomorphic maps in somewhat more general situations.

THEOREM 17. *Let $\phi : \Sigma \to \mathbf{S}^2$ be a non-constant holomorphic map defined on a compact Riemann surface and C be an equator of $\mathbf{S}^2$ such that $\Sigma - \phi^{-1}(C)$ has m connected components $\Omega_1, \ldots, \Omega_m$. If, for each $i = 1, \ldots, m$, the branching values of $\phi_{|\Omega_i}$ lie in another equator C_i of $\mathbf{S}^2$ orthogonal to C, then we have*

$$\text{Ind}\,\phi \geq m-1 \quad \text{Ind}\,\phi + \text{Nul}\,\phi \leq 4\deg\phi + 2 - m.$$

PROOF: For each connected component Ω_i, $i = 1, \ldots, m$, of $\Sigma - \phi^{-1}(C)$, we will consider the compact Riemann surface Σ_i constructed by glueing two isometric copies of Ω_i along their boundaries and the involutive isometry S_i of Σ_i which permutes both copies. Define a holomorphic map $\phi_i : \Sigma_i \to \mathbf{S}^2$ by extending $\phi_{|\Omega_i}$ by means of the symmetry S_i. So, we have that $\phi_i \circ S_i = S \circ \phi_i$, where S is the symmetry of the two-sphere with respect to the equator C. Moreover, from our hypothesis, all the branching values of ϕ_i lie in the equator C_i. If we denote by d_i the degree of ϕ_i, then $2\deg\phi = d_1 + \cdots + d_m$ and, from Corollary 15,

$$\text{Nul}\,\phi_i = 3 \qquad \text{Ind}\,\phi_i = 2d_i - 1.$$

Since each Ω_i was defined as a nodal domain of a linear function of ϕ, the first eigenvalue of the operator L_ϕ, see (1-8), for the Dirichlet problem on Ω_i is equal to zero. Now, let $f \in V_\lambda(\phi_i)$, see (1-9), a λ-eigenfunction of L_{ϕ_i} on Σ_i which is antiinvariant by the involution S_i. Then f must vanish on the boundary of Ω_i. But we have seen that L_ϕ has no negative eigenvalues on Ω_i. So, we conclude that each eigenfunction in $V_\lambda(\phi_i)$ with $\lambda < 0$ is invariant with respect to S_i. Hence $V_\lambda(\phi_i)_{|\Omega_i}$ consists of eigenfunctions for the Neumann problem of L_ϕ on Ω_i. Coversely, each such eigenfunction on Ω_i, extended via S_i on the whole of Σ_i lies in some of the $V_\lambda(\phi_i)$. As a conclusion there are exactly $2d_i - 1$ eigenfunctions of L_ϕ for the Neumann problem on Ω_i associated to negative eigenvalues. An analogous reasoning shows us that the multiplicity of the zero eigenvalue for the Neumann problem of L_ϕ on Ω_i is two. Now, the proof of this theorem follows directly from Lemmae 12 and 13 putting $k = 1$ and taking into account that $L_\phi = \Delta_\phi + 2$.

The genus zero case. From now on, we will assume that our surface Σ has genus zero, that is, Σ is conformally equivalent to $\overline{\mathbf{C}}$. Let $\phi : \overline{\mathbf{C}} \to \mathbf{S}^2$ be a non-constant holomorphic map and put $d = \deg\phi \geq 1$. If $B(\phi) = m_1p_1 + \cdots + m_kp_k$, $m_1 + \cdots + m_k = 2d - 2$, is the ramification divisor of ϕ, we have, from Theorem 5 and the fact that $H_1(\overline{\mathbf{C}}, \mathbf{Z}) = 0$, that

$$\text{(3-15)} \qquad \frac{N(\phi)}{L(\phi)} \cong \left\{\sigma \in H^0(2\kappa_{\overline{\mathbf{C}}} \otimes [\sum_{i=1}^{k} p_i]) \,|\, \text{Res}_{p_i} \frac{\sigma}{dg} = 0\right\},$$

where $g : \overline{\mathbf{C}} \to \overline{\mathbf{C}}$ is the meromorphic function associated to ϕ in (3-1). The following results are consequences of this isomorphism.

PROPOSITION 18. *For each holomorphic map $\phi : \overline{\mathbf{C}} \to \mathbf{S}^2$, we have* $\text{Nul}\,(\phi) = 3 + 2k$ *for some integer k.*

PROOF: It is sufficient to observe that $N(\phi)/L(\phi)$ is a complex linear space according to (3-15).

THEOREM 19. *Let $\phi : \overline{\mathbf{C}} \to \mathbf{S}^2$ be a holomorphic map. Then*

$$\text{Nul}\,(A \circ \phi) = \text{Nul}\,\phi \quad \text{Ind}\,(A \circ \phi) = \text{Ind}\,\phi$$

for each conformal transformation A of $\mathbf{S}^2$.

PROOF: If g is the meromorphic function associated to ϕ, then the corresponding meromorphic function h associated to $A \circ \phi$ is given by

$$h = \frac{\alpha g + \beta}{\gamma g + \delta} \text{ with } \alpha, \beta, \gamma, \delta \in \mathbf{C} \text{ and } \alpha\delta - \beta\gamma = 1.$$

The ramification divisors of ϕ and $A \circ \phi$ coincide. So, if $\sigma \in H^0\left(2\kappa_{\overline{\mathbf{C}}} \otimes \left[\sum_{i=1}^k p_i\right]\right)$ with $\text{Res}_{p_i}(\sigma/dg) = 0$, we have

$$\text{Res}_{p_i} \frac{\sigma}{dh} = \text{Res}_{p_i} \frac{\sigma(\gamma g + \delta)^2}{dg} = 0$$

because $\text{Res}_{p_i}(\sigma g^j/dg) = 0$, $j = 1, \ldots, m_i$, since $g^{j)}(p_i) = 0$. From (3-15), we conclude that

$$\frac{N(\phi)}{L(\phi)} \cong \frac{N(A \circ \phi)}{L(A \circ \phi)}$$

and, so, Nul ϕ =Nul$(A \circ \phi)$. With respect to the index, as the Lie group $\mathcal{C}(\mathbf{S}^2)$ of the conformal transformations of $\mathbf{S}^2$ is connected, we can take a continuous curve $A_t \in \mathcal{C}(\mathbf{S}^2$, $t \in [0,1]$, with $A_0 = I$ and $A_1 = A$. Consider now the continuous family $Q_t = Q_{A_t \circ \phi}$ of quadratic forms on $W_1(\overline{\mathbf{C}})$ defined in (1-2). As we already know that the nullity of Q_t has constant dimension and since the eigenvalues of Q_t depend continuously on t, we get that index Q_t is constant. In particular, Ind ϕ = index Q_0 = index Q_1 = Ind $(A \circ \phi)$.

From this last invariance theorem, we will be able to obtain a lower bound for the index of a holomorphic map defined on $\overline{\mathbf{C}}$. Firstly, we need to prove the following lemma.

LEMMA 20. *Given an integer $d \geq 1$ and a point $a \in \mathbf{S}^2$, there exists a real number $\varepsilon > 0$ such that, if $\phi : \Sigma \to \mathbf{S}^2$ is a holomorphic map on a compact Riemann surface with $\deg \phi = d$ satisfying:*

a) $\#\phi^{-1}(a) = k$,

b) *the branching values of ϕ different from a are at distance less than ε from $-a$,*

then

$$\text{Ind}\,\phi \geq 2d - k.$$

PROOF: Let $d_1, \ldots, d_k$ be the multiplicities of ϕ at the points of $\phi^{-1}(a)$. Take a number $0 < r < \pi$ and denote by $\Omega_r \subset \Sigma$ the inverse image by ϕ of a geodesic disk of $\mathbf{S}^2$ with radius $\pi - r$ and center a. Then, the open set Ω_r is isometric to the disjoint union of geodesic disks of radius $\pi - r$ and center at the origin of $(\overline{\mathbf{C}}, ds_{\phi_i^2})$, where $\phi_i : \overline{\mathbf{C}} \to \mathbf{S}^2$ is given by $z \mapsto z^{d_i}$, . From standard properties of eigenvalues, we have that Ind ϕ is greater than or equal to the number of negative eigenvalues of L_ϕ for the Dirichlet problem on Ω_r. When r goes to zero this number gets close to

$$\sum_{i=1}^k \text{Ind}\,\phi_i = \sum_{i=1}^k (2d_i - 1) = 2d - k,$$

from Corollary 15.

THEOREM 21. *Let $\phi : \overline{\mathbf{C}} \to \mathbf{S}^2$ be a non–constant holomorphic map. If m is the greatest ramification order of ϕ, then*

$$\text{Ind } \phi \geq m + \deg \phi.$$

So, if $\deg \phi > 1$, we have that

$$\text{Ind } \phi \geq 1 + \deg \phi$$

and, if ϕ has a ramification point with order $\deg \phi - 1$, then

$$\text{Ind } \phi \geq 2 \deg \phi - 1.$$

PROOF: Given a ramification point $p \in \overline{\mathbf{C}}$ with order m, we choose an 1-parameter subgroup F_t, $t \in \mathbf{R}$, of conformal transformations of the two–sphere such that

$$F_t(\phi(p)) = \phi(p) \quad \text{for all } t \in \mathbf{R}$$

and

$$\lim_{t\to\infty} F_t(q) = -\phi(p) \quad \text{for all } q \in \mathbf{S}^2 - \{\phi(p)\}.$$

Then, for great enough $t \in \mathbf{R}$, $F_t \circ \phi : \overline{\mathbf{C}} \to \mathbf{S}^2$ is a holomorphic map with $\deg(F_t \circ \phi) = \deg \phi$ and such that all its branching values, except $\phi(p)$, are as close as one wants to $-\phi(p)$. So, by using Lemma 20 above and Theorem 19, we obtain

$$m + \deg \phi \leq 2 \deg \phi - \#\phi^{-1}(\phi(p)) \leq \text{Ind } (F_t \circ \phi) = \text{Ind } \phi.$$

COROLLARY 22. *Let $\phi : \overline{\mathbf{C}} \to \mathbf{S}^2$ be a holomorphic map. Then*

a) Ind $\phi = 0$ *if and only if ϕ is constant.*
b) Ind $\phi = 1$ *if and only if* $\deg \phi = 1$.
c) *The case* Ind $\phi = 2$ *is not possible.*
d) Ind $\phi = 3$ *if and only if* $\deg \phi = 2$.

PROOF: The statements a), b) and c) are already known, see [**Cho**] and [**N**], and follow easily from our Theorem 21. If we have Ind $\phi = 3$, then $1 + \deg \phi \leq 3$ and, so, $\deg \phi = 2$. Conversely, if $\deg \phi = 2$, then ϕ has only two branching values and we can apply Corollary 15.

Now we will return to consider the isomorphism (3-15) which will allow us to achieve a satisfactory description of the nullity and, as a consequence, of the index of holomorphic maps defined on the completed complex plane. Let $g : \mathbf{C} \to \mathbf{C}$ be a meromorphic function with degree $d > 1$ and $z_1, \ldots, z_{2d-2}$ its ramification points which we will suppose all with order one. Moreover, without lost of generality, we may assume that $z_i \neq \infty$ and $g(z_i) \neq \infty$, $i = 1, \ldots, 2d-2$. If $\phi_g : \overline{\mathbf{C}} \to \mathbf{S}^2$ is the holomorphic map corresponding to g, from the aforementioned isomorphism (3-15), we have that the linear space $N(\phi_g)/L(\phi_g)$ can be described in terms of the space of meromorphic quadratic differentials on $\overline{\mathbf{C}}$ with simple poles at the points $z_1, \ldots, z_{2d-2}$. Such a differential σ can be written as follows

$$\sigma = \sum_{i=1}^{2d-2} \frac{A_i}{z - z_i}(dz)^2 = \frac{R(z)}{(z - z_1)\cdots(z - z_{2d-2})}(dz)^2 \tag{3-16}$$

where $A_i \in \mathbf{C}$ and R is a polynomial with degree less than or equal to $2d-6$ (in particular, notice that Nul $\phi = 3$ if $\deg\phi = 1, 2$), that is

$$\begin{aligned} &\sum_{i=1}^{2d-2} A_i = 0 \\ (3\text{-}17) \qquad &\sum_{i=1}^{2d-2} S_1(z_1,\ldots,\widehat{z}_i,\ldots,z_{2d-2})A_i = 0 \\ &\sum_{i=1}^{2d-2} S_2(z_1,\ldots,\widehat{z}_i,\ldots,z_{2d-2})A_i = 0 \end{aligned}$$

where S_1 and S_2 are the first and second elemental symmetric polynomials in the corresponding variables. Now, we will compute the residues of the meromorphic differential σ/dg at its two–order poles z_i:

$$\mathrm{Res}_{z=z_k}\frac{\sigma}{dg} = \frac{d}{dz}\bigg|_{z=z_k}\left\{\frac{(z-z_k)^2}{g'(z)}\sum_{i=1}^{2d-2}\frac{A_i}{z-z_i}\right\} = \frac{1}{g''(z_k)}\sum_{i\neq k}\frac{A_i}{z_k-z_i} - \frac{1}{2}\frac{g'''(z_k)}{g''(z_k)}A_k.$$

Hence, the diffrential σ/dg has no residues if and only if

$$(3\text{-}18) \qquad \sum_{i\neq k}\frac{A_i}{z_k-z_i} - \frac{1}{2}\frac{g'''(z_k)}{g''(z_k)}A_k = 0, \quad k = 1,\ldots,2d-2.$$

Notice that the system of equations (S_g) given by (3-17) and (3-18) with $A_1,\ldots,A_{2d-2}$ as unknowns determine Nul ϕ_g in the following way:

$$(3\text{-}19) \qquad \mathrm{Nul}\;\phi_g = 3 + 2(2d-2) - 2r(g)$$

where $r(g)$ is the rank of the complex $(2d+1)\times(2d-2)$-matrix $Z(g)$ of the coefficoents of the system (S_g).

On the other hand, let M_d be the space of all the meromorphic functions with degree d defined on $\overline{\mathbf{C}}$. There exists a natural action of the group $\mathcal{C}(\mathbf{S}^2)$ of the conformal transformations of $\mathbf{S}^2 \equiv \overline{\mathbf{C}}$ on M_d given by

$$(A, g) \in \mathcal{C}(\mathbf{S}^2)\times M_d \longmapsto A\circ g \in M_d.$$

The corresponding quotient space $\mathcal{M}_d = M_d/\mathcal{C}(\mathbf{S}^2)$ was identified in [**Loo**] with the complement of an algebraic hypersurface in the complex Grassmannian $G(2, d+1)$ by

$$[g = \frac{P}{Q}] \in \mathcal{M}_d \longmapsto \mathrm{span}\{P, Q\}$$

where P, Q are polynomials of degree d without common roots. Moreover the ramification map $\psi_d : \mathcal{M}_d \to \mathbf{P}^{2d-2}$ given by

$$\psi_d\left(\left[\frac{P}{Q}\right]\right) = \left[\operatorname{coeff}(P'Q - Q'P)\right]$$

is shown in [**Loo**] to be a branched covering with $(2d-2)!/(d-1)!d!$ sheets. Consider another branched covering $\rho_d : \mathbf{C}^{2d-2} \to \mathbf{P}^{2d-2}$ defined by

$$\rho_d(z_1, \ldots, z_{2d-2}) = \left[\operatorname{coeff}(z - z_1)\cdots(z - z_{2d-2})\right].$$

So, except on an algebraic hypersurface $\mathcal{H}$, we may introduce on $\mathcal{M}_d$ local complex coordinates by means of the multivalued map $\psi_d^{-1} \circ \rho_d$. Observe that, from Theorem 19, we may consider Nul and Ind as integer valued functions on the space $\mathcal{M}_d$ and that, if z is a ramification point of a meromorphic function g and $A \in \mathcal{C}(\mathbf{S}^2)$, then

$$\frac{(A \circ g)'''(z)}{(A \circ g)''(z)} = \frac{g'''(z)}{g''(z)}.$$

Hence, the map given by

$$[g] \in \mathcal{M}_d - \mathcal{H} \longmapsto Z(g)$$

is analytic. Now, we can state

THEOREM 23. *The set $\mathcal{N}_d$ consisting of the equivalence classes of meromorphic functions of degree d in $\mathcal{M}_d$ where* Nul > 3 *is an analytic subset. Then, we have* Ind $= 2d-1$ *on $\mathcal{M}_d - \mathcal{N}_d$ and* Ind $\leq 2d-1$ *everywhere on $\mathcal{M}_d$.*

PROOF: We know that the map $[g] \mapsto Z(g)$ is analytic and, from (3-19), we have

$$\mathcal{N}_d \subset \mathcal{R} = \{[g] \in \mathcal{M}_d - \mathcal{H} \,|\, r(g) < 2d-2\} \cup \mathcal{H}.$$

So, $\mathcal{N}_d$ will be analytic if we prove that the analytic set $\mathcal{R}$ does not coincide with $\mathcal{M}_d$. In fact, we can take a holomorphic map $g_0 : \overline{\mathbf{C}} \to \mathbf{S}^2$ with degree d and such that all its branching values lie in an equator of $\mathbf{S}^2$. From Corollary 15 and (3-19) and by remembering that the eigenvalues of L_{ϕ_g} depend continuously on g, we have that there are points in a open neighbourhood of $[g_0]$ which lie in $\mathcal{M}_d - \mathcal{R}$.

Hence, the set $\mathcal{N}_d$ does not disconnect the space $\mathcal{M}_d$ and, so, all the points in its complement must have the same index. In fact, one can join any two of them by a continuous curve and when one moves along it the nullity is unchanged. Then, from the continuity of the eigenvalues, the index does not change as well. But, from Corollary 15 again, there are points in that complement whose index is exactly $2d-1$. As a conclusion, Ind is identically $2d-1$ on $\mathcal{M}_d - \mathcal{N}_d$. If one uses again the continuity of the eigenvalues, one has that Ind $\leq 2d-1$ on the whole of $\mathcal{M}_d$.

So, in the genus zero case, we have obtained an upper bound for the index which cannot be improved because it is attained by almost all meromorphic functions. This last theorem has been obtained independently by Ejiri and Kotani [**E–Ko**]. Combining this upper bound with Theorem 21, we have

COROLLARY 24. *Let $\phi : \overline{\mathbf{C}} \to \mathbf{S}^2$ be anon-constant holomorphic map which has a ramification point with order $\deg\phi - 1$ (for instance, $\phi(z) = a_r z^r + \cdots + a_1 z + a_0$). Then*

$$\operatorname{Ind}\phi = 2\deg\phi - 1.$$

Finally, we will do a detailed study of the degree three case because there one already gets some anomalies which point out that the explicit formula asked for by Fischer-Colbrie and Gulliver does not exist probably. In fact, we will show that all the meromorphic functions of degree three on $\overline{\mathbf{C}}$ behave in the generic way described in Theorem 23, except in a particular situation: when its four ramification points form an equianharmonic quadruple.

Let $\phi : \overline{\mathbf{C}} \to \mathbf{S}^2$ be a degree three holomorphic map and $z_1, z_2, z_3, z_4 \in \overline{\mathbf{C}}$ its ramification points. If any two of them coincide, then their images (at most three) can be placed on an equator of $\mathbf{S}^2$ by using a conformal transformation. So, in this case, from Corollary 15, we have $\operatorname{Nul}\phi = 3$ and $\operatorname{Ind}\phi = 5$. Suppose now that z_1, z_2, z_3, z_4 are different points of $\overline{\mathbf{C}}$. In that case, the space

$$H^0\left(2\kappa_{\overline{\mathbf{C}}} \otimes [z_1 + z_2 + z_3 + z_4]\right)$$

of the meromorphic quadratic differentials on $\overline{\mathbf{C}}$ with simple poles at z_1, z_2, z_3, z_4 is a complex 1-dimensional space. In fact, the equation (3-16) says us in this case that these differentials are of the form

$$\sigma = \frac{\lambda}{P'Q - Q'P}(dz)^2, \quad \lambda \in \mathbf{C}$$

if P/Q represents the meromorphic function associated to ϕ with $\max\{\deg p, \deg Q\} = 3$. Then, taking into account (3-15), $\operatorname{Nul}\phi > 3$ if and only if $\operatorname{Nul}\phi = 5$ and this occurs if and only if

$$\operatorname{Res}_{z=z_i} \frac{Q^2}{(P'Q - Q'P)^2}\, dz = 0, \quad i = 1, 2, 3, 4.$$

This condition can be easily rewritten as

$$P'''Q - PQ''' = P''Q' - P'Q'',$$

that is, the cross ratio of the four roots z_1, z_2, z_3, z_4 of $P'Q - PQ'$ satisfies

$$(z_1, z_2, z_3, z_4) = \rho \quad \text{with } \rho^2 - \rho + 1 = 0.$$

This is equivalent to the fact that, up to a Möbius transformation, $z_1 = 0$, $z_2 = 1$, $z_3 = \varepsilon$, $z_4 = \varepsilon^2$ where ε is a primitive cubic root of the unity. It is easy to see that, up to conformal transformations, the meromorphic function associated to such a ϕ is given by $g(z) = z/(z^3 + 2)$.

So, we have proved that a holomorphic map $\phi : \overline{\mathbf{C}} \to \mathbf{S}^2$ with degree three has $\operatorname{Nul}\phi > 3$ if and only if its ramification points form an equianharmonic quadruple and, in this case, $\operatorname{Nul}\phi = 5$. As conclusions, we can ennonce the following results.

THEOREM 25. *Let* $\phi : \overline{\mathbb{C}} \to \mathbf{S}^2$ *be a degree three holomorphic map. If its four ramification points form an equianharmonic quadruple, then* Nul $\phi = 5$ *and* Ind $\phi = 4$. *Otherwise* Nul $\phi = 3$ *and* Ind $\phi = 5$.

PROOF: The assertions about the nullity have been already proved. With respect to the index, notice that, if Nul $\phi = 3$, then, from Theorem 23, Ind $\phi = 5$. In the case Nul $\phi = 5$, we can apply Theorem 17 with $m = 5$ to obtain Ind $\phi \geq 4$ and Ind $\phi +$ Nul $\phi \leq 9$. So, Ind $\phi = 4$.

COROLLARY 26. *Let* $\phi : \overline{\mathbb{C}} \to \mathbf{S}^2$ *be a holomorphic map. Then* Ind $\phi = 4$ *if and only if* $\deg \phi = 3$ *and its ramification points form an equianharmonic quadruple.*

PROOF: If Ind $\phi = 4$ we obtain that ϕ had degree three from Theorem 21 and Theorem 23. So, we conclude from Theorem 25.

REMARK. The planar ends complete minimal surfaces associated to the meromorphic function $g(z) = z/(z^3 + 2)$ by means of Theorem 3 are unbranched and have embedded ends. These surfaces have been described by Bryant [**Br1**] and Rosenberg and Toubiana [**Ro–To**].

SOME PROBLEMS

Finally, we want to deal with some problems about this topic which can be interesting from our point of view.

The first one asks if the upper bound that we have obtained for the index of holomorphic maps on the genus zero surface in Theorem 23 is valid for all genera. With respect to this, we think that Ejiri and Micallef have made some progress, but we do not know exactly their results.

The second one is an inverse spectral question but strongly related to complex analysis, beacuse we have a branched metric ds^2_ϕ for each holomorphic map ϕ from a compact Riemann surface and we want to obtain from the spectrum of its Laplacian all the possible information, see (1-7).

A third problem is to look for properties that a holomorphic map must satisfy in order to have nullity greater than three. The Gauss map of complete minimal embedded surfaces with finite total curvature are in this situation, see remark after Proposition 1. In a earlier version of this work, we had conjectured that these holomorphic maps, in the genus zero case, were precisely the branching points of the ramification map ψ_d used in the proof of Theorem 23. In fact Ejiri [**E**] has solved this conjecture in the affirmative. He shows the following equivalent fact: a holomorphic map on the sphere has nullity greater than three if and only if there is a curve of spherical minimal immersions into the four–sphere which collapses in this map.

The last problem asks if each bounded Jacobi field on a complete minimal surface with finite total curvature comes from a deformation of this surface by means of minimal surfaces of the same type, see Conclusion 1. In the examples that we know this is true.

References

[A] N. Aronszajn, *A unique continuation theorem for solutions of elliptic partial differential equations of second order*, J. Math. Pures Appl. **36** (1957), 235–249.

[Ba–do C] J. L. Barbosa and M. P. do Carmo, *On the size of a stable minimal surface in* $\mathbf{R}^3$, Amer. J. Math. **98** (1976), 515- 528.

[Bl] W. Blaschke, "Vorlesungen über Differentialgeometrie III," Springer, Berlin, 1929.

[Br1] R. L. Bryant, *A duality theorem for Willmore surfaces*, J. Diff. Geom. **20** (1984), 23–53.

[Br2] R. L. Bryant, *Surfaces in conformal geometry*, Proc. of Symp. in Pure Math. **48** (1988), 227–240.

[do C–Pe] M. P. do Carmo and C. K. Peng, *Stable complete minimal surfaces in* $\mathbf{R}^3$ *are planes*, Bull. Am. Math. Soc. **1** (1979), 903–905.

[Che] S. S. Chern, *On minimal spheres in the four sphere*, in "Selected Papers," Springer–Verlag, New York, 1988, pp. 421–434.

[Cho] J. Choe, *Index, vision number and stability of complete minimal surfaces*, Arch. for Rat. Mech. and An. **109** (1990), 195–212.

[Co–Hi] R. Courant and D. Hilbert, "Methods of Mathematical Physics," Interscience, New York, 1953.

[E] N. Ejiri, *Minimal deformations from a holomorphic map of* $\mathbf{S}^2$ *onto* $\mathbf{S}^2(1)$ *to a minimal surface in* $\mathbf{S}^4(1)$, preprint.

[E–Ko] N. Ejiri and M. Kotani, *Index and flat ends of minimal surfaces*, preprint.

[Fa–Kr] H. M. Frakas and I. Kra, "Riemann surfaces," Springer–Verlag, New York, 1979.

[FC] D. Fischer-Colbrie, *On complete minimal surfaces with finite Morse index in three manifolds*, Invent. Math. **82** (1985), 121–132.

[FC–Sch] D. Fischer-Colbrie and R. Schoen, *The structure of complete stable minimal surfaces in three manifolds of non–negative scalar curvature*, Comm. of Pure and Appl. Math. **33** (1980), 199–211.

[Gr–Ha] P. Griffiths and J. Harris, "Principles of algebraic geometry," Wiley Interscience, New York, 1978.

[Gu] R. Gulliver, *Index and total curvature of complete minimal surfaces*, Proc. Symp. Pure Math. **44** (1986), 207–211.

[Gui–Ka] V. Guillemin and D. Kazhdan, *Some inverse spectral results for negatively curved 2-manifolds*, Topology **19** (1980), 301–312.

[He] J. Hersch, *Quatre propiétés isoperimetriques de membranes spheriques homogènes*, C. R. Acad. Sci. Paris **270** (1970), 1645–1648.

[J–M] L. P. Jorge and W. H. Meeks III, *The topology of complete minimal surfaces of finite total Gaussian curvature*, Topology **22** (1983), 203–221.

[La] G. Laffaille, *Déterminants de laplaciens*, Séminaire de Théorie Spectrale et Géometrie (1986), 77–84.

[Li–Tr] P. Li and A. Treibergs, *Applications of eigenvalue techniques to geometry*, preprint.

[Lo] F. J. López, *The classification of complete minimal surfaces with total curvature greater than* -12π, Trans. Amer. Math. Soc. (to appear).

[**Lo–R**] F. J. López and A. Ros, *Complete minimal surfaces with index one and stable mean curvature surfaces*, Comment. Math. Helv. **64** (1989), 34–43.

[**Loo**] B. Loo, *The space of harmonic maps of* $\mathbf{S}^2$ *into* $\mathbf{S}^4$, Trans. Amer. Math. Soc. **313** (1989), 81–102.

[**N**] S. Nayatani, *Lower bounds for the Morse index of complete minimal surfaces in Euclidean 3-space*, preprint.

[**On–V**] E. Onofri and M. A. Virasoro, *On a formulation of Polyakov's string theory with regular classical solutions*, Nucl. Phys. B **201** (1982), 159–175.

[**Os–Ph–Sa**] B. Osgood, R. Phillips and P. Sarnack, *Extremals of determinants of Laplacians*, J. of Funct. An. **80** (1988), 148–211.

[**Oss**] R. Osserman, "A survey of minimal surfaces," Dover, New York, 1986.

[**Po**] A. V. Pogorelov, *On the stability of minimal surfaces*, Soviet Math. Dokl. **24** (1981), 274–276.

[**Ro–To**] H. Rosenberg and E. Toubiana, *Some remarks on deformations of minimal surfaces*, Trans. Amer. Math. Soc. **295** (1986), 491–499.

[**Sch**] R. Schoen, *Uniqueness, symmetry and embeddedness of minimal surfaces*, J. Diff. Geom. **18** (1983), 791–809.

[**Schw**] H. A. Schwarz, "Gesammelte Abhandlungen," Springer, Berlin, 1890, pp. 224–264.

[**Ty**] J. Tysk, *Eigenvalue estimates with applications to minimal surfaces*, Pacific J. of Math. **128** (1987), 361–366.

Departamento de Geometría y Topología. Universidad de Granada. 18071–Granada. Spain

This paper is in final form and no version will appear elsewhere.

Generic existence of Morse functions on infinite dimensional Riemannian manifolds and applications

D. Motreanu

Introduction.

The aim of the paper is to solve the following problem: find conditions implying the (generic) existence of Morse functions in a prescribed family of C^1 functions $\{f_a : M \longrightarrow R\}_{a\in A}$ on a (possibly infinite dimensional) Riemannian manifold. For technical reasons we regard the family as corresponding to a mapping $f : M \times A \longrightarrow R$. Many relevant situations are known where this property holds. For example, if M is a finite dimensional C^k manifold and $A = C^k(M, R)$ with $k \geq 2$, then the Morse functions on M form a dense open set in A (cf. Hirsch [4], p. 147). As another example, if M is an immersed submanifold of $A = R^n$ and $f : M \times A \longrightarrow R$ is given by $f(x, a) = \|x - a\|^2$, then for almost all $a \in A$ the function $f_a = f(\cdot, a) : M \longrightarrow R$ is a Morse function on M (see Milnor [5], p. 36).

Our main results are formulated in Theorem 1.1 and Corollary 1.7 which contain a verifiable criterion, in terms of the second order partial derivatives of f, to have the above mentioned property. From this general result one obtains easily the previous two examples. In addition, Theorem 1.1 supplies also some regularity informations concerning the dependence of the critical points of $f(\cdot, a)$ on the parameter $a \in A$. The study of this dependence was inspired by the work of Saut and Temam [12]. When M is finite dimensional this problem has been treated in [9]. The infinite dimensional case as studied in Theorem 1.1 presents a specific treatment involving the Fredholm operators.

Corallary 1.7 is used in Theorem 2.1 to prove the generic existence of Morse functions on the infinite-dimensional Riemannian manifold of L_1^2-paths joining a fixed point with a compact submanifold. These Morse functions are constructed explicitly as perturbations of the energy integral by Euclidean distance functions.

Finally, by means of a Morse function in Theorem 2.1, one gives in Theorem 3.1 a precise description of a CW-complex having the homotopy type of the preceding path space. When the submanifold A is just a one-point set the result in Theorem 3.1 reduces to the Fundamental Theorem of Morse Theory (see Milnor [5], p. 95). Theorem 3.1 is illustrated by an example treating the space of curves on S^n from a point p to a submanifold diffeomorphic to S^{n-1} not passing through p.

The remaining of the paper is organized as follows. Section 1 contains the abstract result upon the genericity for the existence of Morse functions. Section 2 presents a concrete application of the abstract result in the case of the manifold of paths between a point and a compact submanifold. This is used in Section 3 to provide a geometric description for the cellular structure of the above path space.

I am very grateful to Professor Jean Mawhin for suggesting the infinite-dimensional approach based on Sard-Smale Theorem [14].

1. Existence of Morse functions via transversality

We recall that a property on a topological space is said to be generic if it holds on a residual set, that is, a countable intersection of dense open sets. The next theorem, which is our principal result, establishes that in a parametrized family, the C^k functions with $k \geq 2$, having only nondegenerate critical points exist generically. The finite dimensional version of this result has been given in [9].

Theorem 1.1. Let M be a separable C^k Riemannian manifold, let A be a metrizable and separable, C^k Banach manifold, for $k \geq 2$, and let a function $f : M \times A \longrightarrow R$ satisfy the hypotheses
(H_1) the first partial derivative $f'_x : M \times A \longrightarrow T^*M = TM$, with respect to $x \in M$, exists and f'_x is a C^r mapping with $1 \leq r \leq k$;
(H_2) for every $(x, a) \in M \times A$ such that x is a critical point of $f(\cdot, a) : M \longrightarrow R$, i.e.,

$$f'_x(x, a) = 0_x \in T_xM, \tag{1.1}$$

the second order partial derivative $f"_{xx}(x, a) : T_xM \longrightarrow T_xM$ is a Fredholm operator of index zero;
(H_3) for every (x, a) satisfying (1.1), the following equality holds

$$Ker f"_{xx}(x, a) \cap Ker f"_{xa}(x, a) = 0_x. \tag{1.2}$$

Then the set

$$G = \{a \in A | \, f(\cdot, a) : M \longrightarrow R \text{ has only nondegenerate critical points}\} \tag{1.3}$$

is residual, hence dense, in A. Moreover, for any connected component G_o of G, there exists a countable set $\{g_i\}_{i \in I}$ of C^r mappings from an open neighbourhood of G_o in A into M verifying the formula

$$\{x \in M | \; f'_x(x, a) = 0_x\} = \{g_i(a)\}_{i \in I} \text{ for all } a \in g_o. \tag{1.4}$$

If in addition to $(H_1), (H_2), (H_3)$ it is assumed that
(H_4) for each compact subset A_o of A, the set of points x satisfying (1.1), with $a \in A$, is compact in M, then the set G of (1.3) is also open in A, the mappings $g_i : G_o \longrightarrow M$ are C^r differentiable and each function $f(\cdot, a) : M \longrightarrow R$, with $a \in G$, has finitely many critical points whose number is constant on each component G_o of G.

Remark 1.2. In equality (1.2) the second order partial derivatives $f"_{xx}(x, a)$ and $f"_{xa}(x, a)$ are regarded as operators acting as follows

$$f"_{xx}(x, a) : T_xM \longrightarrow T_xM \text{ and } f"_{xa}(x, a) : T_xM \longrightarrow T^*_aM.$$

Before proceeding to the proof of Theorem 1.1 we present a technical lemma which points out that hypothesis (H_3) reduces to a transversality assertion.

Lemma 1.3. Let the data M, A and f have the same meaning as in Theorem 1.1. Then, for every $(x, a) \in M \times A$ verifying (1.1), the following conditions are equivalent
(i) for each u^* in the vertical space $T_{0^*_x}(T^*_xM) \cong T^*_xM = T_xM$ there exists $(u, w) \in T_xM \times T_aM$ such that

$$f"_{xx}(x, a)(\cdot, u) + f"_{xa}(x, a)(\cdot, w) = u^*; \tag{1.5}$$

(ii) relation (1.2) holds.

Proof. In the case where the manifold M is finite dimensional this lemma has been proved in [9]. Although in the infinite dimensional case the proof follows the same lines like in the finite dimensional setting, we shall outline it for the sake of completeness.

(i) $\Rightarrow$ (ii). The image of the selfadjoint operator $f"_{xx}(x,a) : T_xM \longrightarrow T_x^*M$ is given by

$$imf"_{xx}(x,a) = \{v^* \in T_x^*M \mid v^*(u) = 0 \text{ for all } u \in Kerf"_{xx}(x,a)\} \tag{1.6}$$

Writing (1.5) in the form

$$u^* - f"_{xa}(x,a)(\cdot, w) \in imf"_{xx}(x,a),$$

we see from (1.6) that (1.5) is equivalent to

$$f"_{xa}(x,a)(v,w) = u^*(v) \text{ for all } v \in Kerf"_{xx}(x,a). \tag{1.7}$$

Now, if v is any element of the set in the left hand side of (1.2) and u^* is an arbitrary linear functional of T_x^*M, relation (1.7) implies at once that $v = 0_x$.

(ii) $\Rightarrow$ (i). Hypothesis (H_2) ensures that Ker $f"_{xx}(x,a)$ is a finite dimensional vector subspace of T_xM. Choose a basis $\{e_1,\ldots,e_m\}$ of $Kerf"_{xx}(x,a)$. By the equivalence of relations (1.5) and (1.7) we have to show that, for every $u^* \in T_x^*M$, there is $w \in T_aA$ such that

$$f"_{xa}(x,a)(e_i,w) = u^*(e_i), i = 1,\ldots,m. \tag{1.8}$$

It is an easy matter to deduce from (1.2) the linear independence of $f"_{xa}(x,a)(e_i,\cdot) \in T_a^*A$, $i = 1,\ldots,m$. Then, it turns out that the linear operator

$$w \in T_aA \longrightarrow (f"_{xa}(x,a)(e_i,w))_{i=1,\ldots,m} \in R^m$$

is onto. This proves the existence of some vector $w \in T_aA$ as required in (1.8).

Proof of Theorem 1.1. Consider the mapping $F : M \times A \longrightarrow T^*M = TM$ defined by

$$F(x,a) = f'_x(x,a), \quad (x,a) \in M \times A. \tag{1.9}$$

Hypothesis (H_1) insures that F is C^r-differentiable. Using (H_3) we are in position to apply Lemma 1.3. We thus deduce that the mapping F is transversal to the zero-section

$$N = \{O_x \in T_xM | x \in M\}$$

of TM which is known to be a C^{k-1} submanifold of TM. Hence $V = F^{-1}(N)$ is a C^r submanifold of $M \times A$.

Denote by $P : V \longrightarrow A$ the restriction to V of the natural projection $M \times A \longrightarrow A$. We claim that the next assertions are valid:

$$a \in A \text{ is a regular value of } P. \Leftrightarrow \tag{1.10}$$

$$F(\cdot,a) : M \longrightarrow TM \text{ is transversal to } N.$$

and:

$$P \text{ is a Fredholm map of index zero.} \tag{1.11}$$

The properties in (1.10) and (1.11) beeing of local nature, in order to check them we may suppose that M and A are open subsets in a Hilbert space H and a Banach space

B, respectively. Under this reduction we have

$$N = M \times O \subset M \times H$$

and

$$F(x,a) = (x, f'_x(x,a)) \in M \times H, \quad (x,a) \in M \times A. \tag{1.12}$$

The transversality of F to N implies that $O \in H$ is a regular value of $f'_x = pr_2 F : M \longrightarrow H$. By Lemma A.2 in [12], this fact and hypothesis (H_2) yield (1.11) and the assertion:

$$a \in A \text{ is a regular value of } P : V \longrightarrow A. \Leftrightarrow \tag{1.13}$$

$$O \in H \text{ is a regular value of } pr_2 F(\cdot, a) : M \longrightarrow H.$$

From (1.13) one obtains (1.10), so the claim is proved.

The properties (1.10) and (1.11) allow us to apply the Sard-Smale theorem [14] to the mapping F introduced in (1.9). We thus derive that the set

$$G = \{a \in A | F(\cdot, a) : M \longrightarrow TM \text{ is transveral to } N\} \tag{1.14}$$

is residual in A. By the Baire theorem this set is dense in A. We must show that the sets described in (1.3) and (1.14) coincide. Indeed, one sees that $a \in G$ if and only if $f"_{xx}(x,a) : T_x M \longrightarrow T_{0_x}(T_x M) = T_x M$ is onto for every $x \in M$ solving (1.1), so if and only if $f"_{xx}(x,a)$ is an isomorphism. The last equivalence is due to the fact that, by (H_2), the Fredholm operator $f"_{xx}(x,a)$ has index zero. The equality of the sets (1.3) and (1.14) follows.

Let us fix now a component G_o of G and a point $a_o \in G_o$. Locally, under representation (1.12), equation (1.1) becomes

$$pr_2 F(x,a) = 0. \tag{1.15}$$

Combining (1.10), (1.13), (1.14) and applying the implicit function theorem, we infer the existence of a countable set of C^r mappings $g_i : A \longrightarrow M$, $i \in I$, defined near a_o in A such that

$$pr_2 F(g_i(a), a) = 0. \tag{1.16}$$

Therefore formula (1.4) is satisfied. The uniqueness of the implicit function and the connectedness of G_o assure that the mappings g_i are defined everywhere on G_o.

Add now assumption (H_4). If C is a compact subset of A, then hypothesis (H_4) implies that

$$P^{-1}(C) = \{(x,a) \in M \times A | \ f'_x(x,a) = 0_x, \ a \in C\} \tag{1.17}$$

is a compact subset of V, or in other words, that $P : V \longrightarrow A$ is a proper mapping. This fact and (1.11) permit to invoke a result due to K. Geba [2] ensuring the openness and the density of the regular values of P. From (1.10) and (1.14) one concludes that the dense set G is also open in A. For $a \in G$, the set of critical points of $f(\cdot, a)$, which is just $F(\cdot, a)^{-1}(N)$, is a discrete space. It is so because, in view of relations (1.11) and (1.14), $F(\cdot, a)^{-1}(N)$ is a C^r submanifold in M of dimension

$$dim F(\cdot, a)^{-1}(N) = dim P^{-1}(a) = index P = 0.$$

This set being also compact, as seen from (H_4), it is necessarily finite. Formula (1.4) indicates that the number of critical points of $f(\cdot, a)$ is independent of a when $a \in G$ runs in the same connected component G_o of G. The proof is complete.

Remark 1.4. If all hypotheses (H_1)-(H_4) are satisfied, the final conclusion of Theorem 1.1 holds without the separability assumption for the manifolds M and A. Hypothesis (H_2) is always satisfied in the case where the manifold M is finite dimensional.

Remark 1.5. The parametric transversality theorems (see Abraham-Robbin [1], p. 48, and Quinn [11]) cannot be used in the above proof because these theorems ask some finite dimension assumptions for the manifold M that are not imposed here.

Remark 1.6. Assume (H_1), (H_3) and replace (H_2) by the weaker hypothesis (H_2') for every $(x, a) \in M \times A$ such that (1.1) is verified, the derivative $f''_{xx}(x, a) : T_xM \longrightarrow T_xM$ is a Fredholm operator.
Arguing as in the proof of Theorem 1.1 one obtains that

$$G_1 = \{a \in A |\ f(\cdot, a) : M \longrightarrow R \text{ has all its critical points } x \in M \text{ with} f''_{xx}(x, a) \text{ surjective}\}$$

is a residual set in A. In this case, for every $a \in G_1$, the set of critical points $f(\cdot, a) : M \longrightarrow R$ forms a countable union of C^r submanifolds in M of dimension equal to the index of $f''_{xx}(x, a)$. The sets G and G_1 coincide if and only if (H_2) holds.

We recall that a Morse function on a Riemannian manifold M means a C^1 function $h : M \longrightarrow R$ whose critical points are all nondegenerate and which verifies the Palais-Smale condition: if $\{x_i\}_{i \geq 1}$ is a sequence in M such that $h(x_i)$ is bounded and the sequence of gradients $h'(x_i)$ satisfies $\|h'(x_i)\|_{x_i} \to 0$ as $i \to \infty$, then $\{x_i\}$ contains a convergent subsequence. This condition is an essential tool in the infinite dimensional Morse theory (see, e.g., [3] and [13]). An extension of it, motivated by some minimization problems with nonconvex constraints, can be found in [7]. Theorem 1.1 yields the following generic result for the existence of Morse functions in a prescribed family.

Corollary 1.7. Assume hypotheses (H_1), (H_2), (H_3) hold and, for every $a \in A$, the function $f(\cdot, a) : M \longrightarrow R$ satisfies the Palais-Smale condition. Then

$$G = \{a \in A |\ f(\cdot, a) : M \longrightarrow R \text{ is a Morse function}\} \tag{1.18}$$

is a residual set of A, hence dense in A. The other regularity properties mentioned in Theorem 1.1 are true.

Remark 1.8. Note that if hypothesis (H_4) is verified, then every function $f(\cdot, a) : M \longrightarrow R$, $a \in A$, satisfies the Palais-Smale condition.

2. Morse functions on a path space

This Section is devoted to the construction of a Morse function on the space of curves in a Riemannian manifold joining a point with a submanifold. This construction is based on Theorem 1.1 and Corollary 1.7. In the sequel smooth means C^∞. A study concerning the geodesics from a point to a submanifold can be found in [3] and [8].

Let V be a smooth, complete and connected Riemannian manifold embedded in some Euclidean space R^n. The Riemannian metric of V, denoted by $q \longrightarrow < \cdot, \cdot >_q$, is not necessarily induced by the usual Riemannian metric of R^n that we designate by

$< \cdot, \cdot >$. Let us fix a point $p \in V$ and a smooth submanifold A of V. Throughout $L_1^2(I, V)$ stands for the (Hilbert) complete Riemannian manifold of absolutely continuous maps from the unit interval $I = [0,1]$ to V having square integrable derivative. Under these conditions

$$M = \{x \in L_1^2(I, V) | \; x(0) = p, \; x(1) \in A\} \tag{2.1}$$

is a smooth submanifold of $L_1^2(I, V)$ that we endow with the induced Riemannian structure. The tangent space T_xM of M at $x \in M$ admits the following characterization

$$T_xM = \{X \in L_1^2(I, TV) | \; X(t) \in T_{x(t)}V \text{ for all } t \in I, \; X(0) = 0_p \text{ and } X(1) \in T_{x(1)}A\}. \tag{2.2}$$

According to Theorem 1.1. we consider the smooth mapping $f : M \times A \longrightarrow R$ defined by

$$f(x, a) = \frac{1}{2}[\int_0^1 \|x'(t)\|_{x(t)}^2 dt + \|x(1) - a\|^2], \; (x, a) \in M \times A \tag{2.3}$$

where $\| \cdot \|_q$ denotes the norm on T_qM induced by $< \cdot, \cdot >_q$, while $\| \cdot \|$ designates the Euclidean norm on R^n.

We establish now that generically with respect to the parameter $a \in A$ the functions $f(\cdot, a) : M \longrightarrow R$ of (2.3) are Morse functions. In the following statement the orthognonal complement of a subspace S in R^n is denoted by $S^\perp$.

Theorem 2.1. Assume that the manifold V is compact and the next conditions are verified:
(a) For any points $u, v \in A$ so that $v - u \in (T_uA)^\perp$, one has $T_uA \cap (T_vA)^\perp = \{0\}$;
(b) There is no point $q \in A$ such that p and q be conjugate along a geodesic of V from p to q.
Then, corresponding to the map f of (2.3), the set G in (1.18) is residual in A. Furthermore, the regularity for the dependence of the critical points of $f(\cdot, a) : M \longrightarrow R$ with respect to $a \in G$ as presented in (1.4) is valid.

To prove Theorem 2.1 we need a preliminary result. It holds without the additional assumptions (a) and (b).

Lemma 2.2. If the manifold A is compact, then the function $f(\cdot, a)$ on M given by (2.3) satisfies the Palais-Smale condition for every $a \in A$.

Proof. The argument is essentially the same as that developed by K. Grove [3], Theorem 2.4, to prove the Palais-Smale condition for the energy functional $E : M \longrightarrow R$,

$$E(x) = \frac{1}{2} \int_0^1 \|x'(t)\|_{x(t)}^2 dt, \; x \in M. \tag{2.4}$$

For a fixed point $a \in A$, let $\{x_i\}_{i \geq 1}$ be a sequence in M such that $f(x_i, a)$ is bounded and $\|f_x'(x_i, a)\|_{x_i} \to 0$ as $i \to \infty$. The equality

$$f(x, a) = E(x) + \frac{1}{2}\|x(1) - a\|^2, x \in M, \tag{2.5}$$

combined with the boundedness of A and $\{f(x_i, a)\}_{i \geq 1}$, shows that $E(x_i)$ is bounded. Consequently, we may suppose that the sequence $\{x_i\}_{i \geq 1}$ converges uniformly in M to some $x \in M$. This allows us to consider that each x_i belongs to the domain of the natural chart determined by the exponential map $exp_x : T_xM \longrightarrow M$. We identify x_i with its representative in this chart. One derives from (2.5) and the estimate in [3] for E that there exist constants $C_1 > 0$ and C_2 such that

$$C_1\|x_i - x_j\|_x^2 \leq (f'_x(x_i,a) - f'_x(x_j,a))(x_i - x_j) + \|x_i - x_j\|_\infty^2 \tag{2.6}$$

for all i, j sufficiently large, where $\|\cdot\|_\infty$ represents the C^o-norm on $C(I, R^n)$. It appears directly from inequality (2.6) that $\{x_i\}_{i\geq 1}$ is a Cauchy sequence. This proves the lemma.

Proof of Theorem 2.1. According to Lemma 2.2 and Corollary 1.7 we have to check that (H_1)-(H_3) are true for the function f in (2.3).

Hypothesis (H_1) is clearly verified. From (2.4), (2.5) it follows that, at any $(x, a) \in M \times A$,

$$f'_x(x,a)(X) = \int_0^1 < \nabla_x X(t), x'(t) >_{x(t)} dt + < x(1) - a, X(1) > \tag{2.7}$$

for all $X \in T_xM$, where ∇ denotes the covariant derivative on V. Then (2.1), (2.2) and (2.7) imply the characterization

$$f'_x(x,a) = 0_x \Leftrightarrow x : I \longrightarrow V \text{ is a geodesic on } V \text{ from } p \text{ to } A \text{ with } x(1) - a \in (T_{x(1)}A)^\perp. \tag{2.8}$$

By (2.7) one sees that for any critical point x of $f(\cdot, a)$ the Hessian $f"_{xx}(x,a)$ is given by

$$f"_{xx}(x,a)(X,Y) = \int_0^1 (< \nabla_x X(t), \nabla_x Y(t) >_{x(t)} - < R(X(t), x'(t))x'(t), Y(t) >_{x(t)})dt$$
$$+ < X(1), Y(1) > \text{ for all } X, Y \in T_xM \tag{2.9}$$

(see Grove [3]). Here R denotes the Riemannian curvature tensor of V. The linear map

$$X \in T_xM \longrightarrow \int_0^1 < \nabla_x X(t), \nabla_x \cdot >_{x(t)} dt \in T_x^*M = T_xM \tag{2.10}$$

is obviously an isomorphism. The linear map

$$X \in T_xM \longrightarrow \int_0^1 < R(X(t), x'(t))x'(t), \cdot >_{x(t)} dt \in T_x^*M = T_xM \tag{2.11}$$

is a compact operator. Indeed, for every X, Y belonging to a fixed bounded set in T_xM, one obtains the estimate

$$\sup_{\|Z\|_x \leq 1} \left| \int_0^1 < R(X(t) - Y(t), x'(t))x'(t), Z(t) >_{x(t)} dt \right| \tag{2.12}$$
$$\leq \sup_{\|Z\|_x \leq 1} \left| \int_0^1 \lambda(t)\|X(t) - Y(t)\|_{x(t)}\|Z(t)\|_{x(t)} dt \right| \leq C\|X - Y\|_{L^2(I,R^n)}^2.$$

The function $\lambda(t)$ is an element of $L^2(I, R^n)$ and C is a positive constant depending only on $x \in M$ and the Riemannian structure of V. The compactness of the operator

(2.11) arises from the compactnes of the inclusion $L_1^2 \subset L^2$ and the estimate (2.12). It is routine to observe that the linear operator

$$X \in T_xM \longrightarrow < X(1), \cdot(1) > \in T_x^*M = T_xM \tag{2.13}$$

is compact. Since, by (2.9), the Hessian $f"_{xx}(x,a) : T_xM \longrightarrow T_xM$ is the sum of an isomorphism (2.10) and of the compact operators (2.11), (2.13) it is a Fredholm operator of index zero (cf. Palais [10], p 122). Hence hypothesis (H_2) holds. From (2.7) one obtains

$$f"_{xa}(X,w) = - < X(1), w > \tag{2.14}$$

for all $(X,w) \in T_xM \times T_aA$. Take $X \in Ker f"_{xx}(x,a) \cap Ker f"_{xa}(x,a)$ with (x,a) satisfying (1.1). By (2.2) and (2.14) it turns out that

$$X(1) \in T_{x(1)}A \cap (T_aA)^{\perp}. \tag{2.15}$$

It follows from (2.8), (2.15) and condition (a) that $X(1)$ vanishes. Then, by (2.7), X must belong to the Kernel of the Hessian of the energy integral for L_1^2-curves between p and $x(1)$. A classical result in Milnor [5], p. 78, shows that X is a Jacobi field along the geodesic x. Because X vanishes at the endpoints p and $x(1)$, conditions (b) leads to the conclusion that $X = 0_x \in T_xM$. Therefore hypotheses (H_3) is verified. Corollary 1.7 completes the proof of Theorem 2.1.

3. Application to the homotopy type of a path space

We keep unchanged all the notations of Section 2. The following theorem represents an extension of the Fundamental Theorem of Morse Theory (see Milnor [5] p. 95) which is obtained from our result if the submanifold A reduces to an one-point set.

Theorem 3.1. Suppose that conditions (a) and (b) in the statement of Theorem 2.1 hold. Then the path space M of (2.1) has the homotopy type of a countable CW-complex K. Precisely, K is built as follows: there exists some $a \in A$ such that K possesses one cell of dimension m for each geodesic x of V from p to A with $x(1)$ - a perpendicular on $T_{x(1)}A$ in R^n provided the index of the bilinear form in (2.9) is equal to m.

Proof. Theorem 2.1 ensures the existence of some $a \in A$ with the property that $g = f(\cdot, a) : M \longrightarrow R$ is a Morse function. In fact, such an $a \in A$ runs in a dense subset of A. Since the critical values of a Morse function are isolated, we can choose a strictly increasing sequence $\{r_i\}_{i \geq 1}$ of positive real numbers such that the r_i's are regular values of g and each open interval (r_{i-1}, r_i) contains exactly one critical value of g. Each critical value of g corresponds to finitely many critical points. In the case that p belongs to A we add to $\{r_i\}_{i \geq 1}$, as the first term, some $r_o < 0$. If we denote, for every index i,

$$M_i = \{x \in M | \; g(x) \leq r_i\},$$

it is known that M_i is obtained from M_{i-1} by attaching finitely many handles corresponding to the critical points of g in $g^{-1}(r_{i-1}, r_i)$. This process is explained for example, in the critical neck principle in J.T. Schwartz [13], p. 139. Assuming the existence of a homotopy equivalence h_{i-1} of M_{i-1} with a finite CW-complex K_{i-1}, it follows there exist a finite CW-complex K_i and a homotopy equivalence $h_i : M_i \longrightarrow K_i$

such that $h_i|_{M_{i-1}} = h_{i-1}$. The CW-complex K_i is constructed as an adjunction space by attaching to K_{i-1} a cell for each critical point of g in $g^{-1}(r_{i-1}, r_i)$, the dimension of the cell being equal to the index of the respective critical point. The cellular approximation theorem enables us to modify homotopically the attaching map to get actually a CW-complex K_i containing K_{i-1} as a subcomplex.

Let us define the CW-complex K to be the direct limit of the CW-complexes K_i. Because M is the union of the sets M_i, there is a continuous mapping $h : M \longrightarrow K$ which is determined uniquely by the equations $h|_{M_i} = h_i$ for all i. Theorem 1.3 in Grove [3] shows that the inclusion of M into the space $C(I, 0, 1; V, p, A)$ of continuous paths is a homotopy equivalence. Since, as proved by J. Milnor [6], $C(I, 0, 1; V, p, A)$ is homotopically equivalent to a CW-complex, it follows that M has the homotopy type of a CW-complex. On the other hand it is clear that h induces isomorphisms of all homotopy groups. Therefore Whitehead's theorem can be applied to the mapping $h : M \longrightarrow K$. It results that h is a homotopy equivalence. Notice that, by the characterization given in (2.8) for the critical points of g (which are all of them nondegenerate), the cellular structure of K is the one described in the statement.

Remark 3.2. If we set $A = \{q\}$ (one-point set) in Theorem 3.1, one obtains the Fundamental Theorem of Morse Theory regarding the homotopy type of the space of paths between two nonconjugate points p and q on a finite dimensional complete Riemannian manifold (cf. Milnor [5], p. 95).

We illustrate the results in Theorems 2.1 and 3.1 by an example dealing with a space of paths on the Euclidean sphere S^n.

Example 3.3. Make the following choice in theorems 2.1 and 3.1: $V = S^n$ with $n \geq 2$, $p = (0, \ldots, 0, 1) \in S^n$ and $A = S^{n-1} = \{x = (x_0, \ldots, x_n) \in S^n | x_n = 0\}$. Let us note that, if $u, v \in A$ are so that $v - u$ is orthogonal in R^{n+1} on $T_u A$, then u and v are necessarily antipodal points in S^{n-1}. This readily implies thast condition (a) holds. The position of S^{n-1} relative to p in S^n shows that p is nonconjugate with any point in A with respect to the complete Riemannian manifold V (see Milnor [5], p. 95-96). Condition (ii) is thus satisfied. Therefore Theorems 2.1 and 3.1 apply for the path space M of (2.1) with our specific data V, p, A. The geodesics entering the final part of Theorem 3.1 are the four great circle arcs on S^n from p to a and $-a$ together with those obtained from these ones by adding complete great circles.

References

[1] Abraham, R. ; Robbin, J.: Transversal Mappings and Flows, W.A.Benjamin, New York, 1967.

[2] Geba, K.: The Leray-Schauder degree and framed bordism, in La théorie des points fixes et ses applications a l'analyse, Presses de l'Université de Montréal, Montréal, 1975.

[3] Grove, K.: Condition (C) for the energy intergral on certain path spaces and applications to the theory of geodesics, J. Differential Geometry 8 (1973), 207-223.

[4] Hirsch, M.W.: Differential Topology, Graduate Texts in Mathematics, Springer Verlag, New York, 1976.

[5] Milnor, J.: Morse Theory, Princeton University Press, Princeton, New Jersey, 1963.

[6] Milnor, J.: On spaces having the homotopy type of a CW-complex, Trans. Amer. Math. Soc. 90 (1959), 272-280.

[7] Motreanu, D.: Existence of minimization with nonconvex constraints, J. Math. Anal. Appl. 117 (1986), 128-137.

[8] Motreanu, D.: Tangent vectors to sets in the theory of geodesics, Nagoya Math. J. 106 (1987), 29-47.

[9] Motreanu, D.: Generic existence of Morse functions, Proc. of the Colloq. on Geometry and Topology, Timisoara, 1989, to appear.

[10] Palais, R.S.: Seminar on the Atiyah-Singer Index Theorem, Princeton University Press, Princeton, New Jersey, 1965.

[11] Quinn, F.: Transversal approximation on Banach manifolds, Proc. Symp. Pure Math 15 (1970), 213-222.

[12] Saut, J.C.; Temam, R.: Generic properties of nonlinear boundary value problems, Comm. Partial Differential Equations 4 (1979), 293-319.

[13] Schwartz, J. T.: Nonlinear Functional Analysis, Gordon and Breach Science Publishers Inc., New York, 1969.

[14] Smale, S.: An infinite dimensional version of Sard's theorem, Amer. J. Math. 87 (1965), 861-866.

This paper is in final form and no version will appear elsewhere.

Universitatea "Al.I.Cuza" Iasi
Seminarul Matematic "Al.Myller"
6600 Iasi, Romania

Some extensions of Radon's theorem.

Barbara Opozda

Introduction.

The main object in classical affine differential geometry is the study of non-degenerate hypersurfaces in affine space $\mathbf{R}^{n+1}$ being a homogeneous space under the action of the unimodular affine group $ASL(n+1, \mathbf{R})$. For a given non-degenerate hypersurface of $\mathbf{R}^{n+1}$ there is a unique transversal vector field satisfying some natural conditions, see for instance [B] and [N]. The vector field is an $ASL(n+1, \mathbf{R})$-invariant and is called the affine normal. By using the affine normal every non-degenerate hypersurface can be endowed with a torsion-free connection, a symmetric nondegenerate 2-form and other invariants of the unimodular group. According to the famous idea of F.Klein a fundamental problem is to find a complete set of such invariants. The first results in this respect are due to W. Blaschke and A.P. Norden, see [Sch]. Another fundamental question is that about objects which can be prescribed on a manifold M and conditions which should be fulfilled by the given objects with the aim of realizing M as a hypersurface in $\mathbf{R}^{n+1}$ in such a way that the given objects become the objects induced by the affine normal. A theorem of J. Radon, see e.g. [B], is an answer to the question in case $dim M = 2$. The theorem received a new formulation in [DNV], where also the existence problem for higher-dimensional hypersurfaces was solved.

The aim of the present paper is to state and prove some existence theorems in a more general geometry of hypersurfaces in $\mathbf{R}^{n+1}$. In the theory hypersurfaces are not assumed to be non-degenerate and the induced structure is given by an arbitrary transversal vector field.

1. Preliminaries.

Let M be a hypersurface in $\mathbf{R}^{n+1}$, i.e. M is an n-dimensional connected manifold and an immersion $f : M \longrightarrow \mathbf{R}^{n+1}$ is given. Assume also that M is oriented. Let ξ be a vector field on M transversal to f. Throughout the paper transversal vector fields are assumed to be nowhere vanishing. The standard connection on $\mathbf{R}^{n+1}$ will be denoted by D and the Lie algebra of all tangent vector fields on M by $\mathcal{X}(M)$. We can write the following analogues of Gauss and Weingarten formulas

$$D_X f_* Y = f_* \nabla_X Y + h(X,Y)\xi, \tag{1.1}$$

$$D_X \xi = -f_* SX + \tau(X)\xi \tag{1.2}$$

for $X, Y \in \mathcal{X}(M)$, where $\nabla_X Y$ and SX are tangential to M. It is easily seen that ∇ is a torsion-free connection, h is a symmetric bilinear form, S is a (1,1)-tensor field and τ is a 1-form on M. Of course, all the objects ∇, h, S and τ depend on the choice of ξ. They will be called the objects induced by (f, ξ). For an arbitrary choice of ξ the following equations are satisfied:
Equation of Gauss: The curvature tensor R of ∇ is given by

$$R(X,Y)Z = h(Y,Z)SX - h(X,Z)SY, \tag{1.3}$$

Equation of Ricci:

$$h(X, SY) - h(Y, SX) = 2d\tau(X, Y), \tag{1.4}$$

Equation of Codazzi I:

$$\nabla h(X, Y, Z) - \nabla h(Y, X, Z) = h(X, Z)\tau(Y) - h(Y, Z)\tau(X), \tag{1.5}$$

Equation of Codazzi II:

$$\nabla S(X, Y) - \nabla S(Y, X) = \tau(X)SY - \tau(Y)SX \tag{1.6}$$

for every $X, Y, Z \in \mathcal{X}(M)$. An immediate consequence of the Gauss and Ricci equations is the following formula

$$trR(X, Y) = 2d\tau(Y, X). \tag{1.7}$$

Recall that in the Blaschke geometry f is non-degenerate, $\tau = 0$ on M and the apolarity condition is satisfied: $tr_h \nabla h(X, \cdot, \cdot) = 0$ for every X, where tr_h denotes the trace with respect h. Then, in particular, ∇h and ∇S are symmetric. For the general case we define

$$Q(X, Y, Z) = \nabla h(X, Y, Z) - \nabla h(Y, X, Z). \tag{1.8}$$

By using the first Codazzi equation and (1.7) one can easily verify the following identity

$$\sigma_{XYZ} \nabla Q(X, Y, Z, W) = \sigma_{XYZ} h(Y, W) trR(Z, X) \tag{1.9}$$

for $X, Y, Z, W \in \mathcal{X}(M)$ and where σ_{XYZ} denotes the cyclic permutation sum with respect to X, Y, Z. If $\tau = 0$ on M, then both sides of (1.9) vanish. Let P be the bundle of all linear orientation-preserving frames on M. The bundle projection of P onto M will be denoted by π. For a fixed transversal vector field ξ we define $F : P \longrightarrow ASL(n+1, \mathbf{R})$ by

$$F(l) = (d_x f \circ l, \xi_x, f_x) \tag{1.10}$$

where $x = \pi(l)$. Throughout the paper we shall use the index range

$$1 \leq i, j \leq n+2, \quad 1 \leq \alpha, \beta \leq n.$$

We set

$$(\omega^i_j) = \omega = F^* \tilde{\omega}, \tag{1.11}$$

where $\tilde{\omega}$ is the Maurer-Cartan form on $AGL(n+1, \mathbf{R})$. If we denote the i-th row of ω by ω^i and the j-th column by ω_j, then we get

$$\begin{aligned} (1.12) \qquad & \omega^{n+2} = 0, \\ & (\omega_{n+2})_l(X) = l^{-1}(\pi_* X) \in \mathbf{R}^n \times \{0\} \times \{0\} \subset \mathbf{R}^{n+2}, \\ & (\omega_{n+1})_l(X) = (-l^{-1} S\pi_* X, \tau(\pi_* X)) \in \mathbf{R}^n \times \mathbf{R} \times \{0\} \subset \mathbf{R}^{n+2}, \\ & (\omega^{n+1}_\alpha)_l(X) = h(\pi_* X, e_\alpha), \text{ where } l = (e_1, \ldots, e_n), \\ & (\omega^\beta_\alpha) - \text{ the connection form of } \nabla. \end{aligned}$$

Formulas (1.12) mean that $F^*\tilde{\omega}$ is determined by ∇, h, S and τ. Recall the following theorem, see [Gr]:
Let F_1 and F_2 be two smooth mappings of a connected manifold N into a Lie group G. Then $F_2 = AF_1$ for some $A \in G$ iff $F_1^*\tilde{\omega} = F_2^*\tilde{\omega}$, where $\tilde{\omega}$ is the Maurer-Cartan form on G.
As an immediate consequence of this theorem and formulas (1.12) we get
Theorem 1.1. Let $f^1, f^2 : M \longrightarrow \mathbf{R}^{n+1}$ be two hypersurfaces. If there are vector fields ξ^1, ξ^2 transversal to f^1 and f^2 respectively, for which the following equalities are satisfied

$$\nabla^1 = \nabla^2,\ h^1 = h^2,\ S^1 = S^2,\ \tau^1 = \tau^2,$$

where ∇^1, h^1, S^1, τ^1 and ∇^2, h^2, S^2, τ^2 are the objects induced by (f^1, ξ^1) and (f^2, ξ^2) respectively, then $f^2 = A \circ f^1$ for some $A \in AGL(n+1, \mathbf{R})$.

Some consequences of Theorem 1.1. are given in [0]. One of them is the following
Theorem 1.2. Let $f^1, f^2 : M \longrightarrow \mathbf{R}^{n+1}$ be two hypersurfaces and $rk\, h^1 > 1$ everywhere on M. If there are vector fields ξ^1, ξ^2 transversal to f^1 and f^2 respectively, for which $\nabla^1 = \nabla^2$ and $h^1 = h^2$, then $f^2 = A \circ f^1$ for some $A \in AGL(n+1, \mathbf{R})$.

Notice that ∇ and h are invariants of the second order with respect to $AGL(n+1, \mathbf{R})$, while S and τ are the ones of the third order.

2. Existence theorems.

At first we shall prove the following basic result.
Theorem 2.1. Let M be a connected simply connected n-dimensional manifold endowed with a torsion-free connection ∇, a symmetric bilinear form h, a (1,1)-tensor field S and a 1-form τ satisfying equations (1.3)-(1.6). Then there is an immersion $f : M \longrightarrow \mathbf{R}^{n+1}$ and a vector field ξ transversal to f such that ∇, h, S and τ are the objects induced by (f, ξ).

Proof. The main tool in the proof is the following theorem of P. Griffiths.
Let N be a manifold, G a Lie group with Lie algebra $\underline{g}$ and ω a $\underline{g}$-valued 1-form on N satisfying the Maurer-Cartan equation:

$$d\omega = -\frac{1}{2}[\omega, \omega], \tag{2.1}$$

where $[\cdot,\cdot]$ is the Lie bracket in the Lie algebra $\underline{g}$. Then for every $x \in N$ there is a neighbourhood U of x and a mapping $F : U \longrightarrow G$ such that $F^*\tilde{\omega} = \omega$ on U, where $\tilde{\omega}$ is the Maurer-Cartan form on G.
Choose an orientation on M and take the bundle P as in section 1. By using the objects ∇, h, S, τ we define an $\underline{agl}(n+1, \mathbf{R})$-valued 1-form $\omega = (\omega^i_j)$ by formulas (1.12). A straightforward verification shows that equations (1.3)-(1.6) yield (2.1) for the standard Lie bracket in $\underline{agl}(n+1, \mathbf{R})$. Hence for every $l \in P$ there is a neighbourhood U of l in P and $F : U \longrightarrow AGL(n+1, \mathbf{R})$ such that $F^*\tilde{\omega} = \omega$, where $\tilde{\omega}$ is the Maurer-Cartan form on $AGL(n+1, \mathbf{R})$. As in section 1, π will denote the bundle projection of P onto M.

Let $F(l) = (\tilde{F}(l), \tilde{f}(l))$, where $\tilde{F}(l)$ is the linear part of $F(l)$ and $\tilde{f}(l)$ is the translation part. Since $F^*\tilde{\omega} = \omega$, we have

$$\tilde{F}(l)^{-1}(\tilde{f}_*(Y)) = l^{-1}(\pi_*(Y)) \tag{2.2}$$

for $Y \in T_l P$ and $l \in U$. It means, in particular, that $\tilde{f}_*(Y) = 0$ for vertical Y, i.e. $\tilde{f}(l)$ depends only on $\pi(l)$. Hence there is a mapping $f : \pi(U) \longrightarrow \mathbf{R}^{n+1}$ such that $\tilde{f} = f \circ \pi$ on U. By (2.2) we also have

$$\tilde{F}(l)^{-1} \circ d_x f = l^{-1} \tag{2.3}$$

for $x = \pi(l)$. It follows that $\tilde{F}(l)$ sends the space $\mathbf{R}^n \times \{0\} \subset \mathbf{R}^{n+1}$ onto $d_x f(T_x M)$ and

$$\tilde{F}(l)_{|\mathbf{R}^n \times \{0\}} = d_x f \circ l. \tag{2.4}$$

In particular, f is an immersion. Denote by $\tilde{\xi}(l)$ the last column of $\tilde{F}(l)$. Then

$$\tilde{F}(l)^{-1}(\tilde{\xi}_*(Y)) = (-l^{-1}(S\pi_*(Y)), \tau(\pi_*(Y))), \tag{2.5}$$

i.e. $\tilde{\xi}(l)$ depends only on $\pi(l)$. Hence there is a vector field $\xi : \pi(U) \longrightarrow \mathbf{R}^{n+1}$ such that $\tilde{\xi} = \xi \circ \pi$. It is clear that ξ is transversal to f. Moreover, we have $F(l) = (d_x f \circ l, \xi_x, f_x)$, where $x = \pi(l)$, i.e. F is given by (1.10). Since $F^*\tilde{\omega} = \omega$, formulas (1.12) imply that ∇, h, S, τ are the objects induced by (f, ξ). By the simply connectedness of M and by Theorem 1.1, such an immersion f can be found globally on M. The proof is completed.

Theorem 1.2. suggests that under additional assumptions we should be able to realize a manifold M as a hypersurface on $\mathbf{R}^{n+1}$ if the given data are ∇ and h only. The classical theorem of Radon is an example of such a result. More precisely, in the theorem of Radon the given data are h and a cubic form C, but ∇ can be easily obained from C and h. The following theorems are some extensions of Radon's theorem as well as the results of [DNV]. The cases where $dim M > 2$ and $dim M = 2$ will be considered separately. In the case of $dim M > 2$ we have

Theorem 2.2. Let M be a connected simply connected n-dimensional manifold, $n > 2$, equipped with a torsion-free connection ∇ and a symmetric bilinear form h with $rk\, h > 2$ everywhere on M. If the following conditions are satisfied:

i) $R(X,Y)Z = 0$ for every X, Y, Z mutually h-orthogonal,

ii) $Q(X,Y,Z) = 0$ for every X, Y, Z mutually h-orthogonal,

iii) $\sigma_{XYZ} \nabla Q(X,Y,Z,W) = \sigma_{XYZ} h(Y,W) tr R(Z,X)$

for every $X, Y, Z, W \in \mathcal{X}(M)$, then there is an immersion $f : M \longrightarrow \mathbf{R}^{n+1}$ and a vector field ξ transversal to f such that ∇ and h are induced by (f, ξ).

Proof. Let $x \in M$ and let $e_1, \ldots, e_n$ be a basis of $T_x M$ adapted to h, i.e.

$$h(e_i, e_j) = \begin{cases} 0 & \text{for } i \neq j \\ 1, -1 \text{ or } 0 & \text{for } i = j, \end{cases}$$

and the first places in the sequence $e_1, \ldots, e_n$ take the vectors for which $h(e_i, e_i) \neq 0$. We set $\varepsilon_i = h(e_i, e_i)$. Let i, j, k be mutually distinct indices. Assume also that $\varepsilon_j \neq 0$ and $\varepsilon_k \neq 0$. Then the vectors $e_i, e_j + e_k, \varepsilon_j e_j - \varepsilon_k e_k$ are mutually h-orthogonal and by i)

$$R(e_i, e_j + e_k)(\varepsilon_j e_j - \varepsilon_k e_k) = 0.$$

Therefore, by using also i) we get

$$\varepsilon_j R(e_i, e_j)e_j = \varepsilon_k R(e_i, e_k)e_k. \tag{2.6}$$

Since $rk\, h > 1$, for every i there is $j \neq i$ such that $\varepsilon_j \neq 0$. We define S by

$$Se_i = \varepsilon_j R(e_i, e_j)e_j. \tag{2.7}$$

By formula (2.6) we see that the definition does not depend on the choice of j. It is easy to observe that S satisfies the Gauss equation. Indeed it is sufficient to check the equation for X, Y, Z which are arbitrary vectors of the given basis. Let $X = e_i$, $Y = e_j$, $Z = e_k$. If i, j, k are mutually distinct or if $i \neq j$, $j = k$ and $\varepsilon_j = 0$, then by i) both sides of (1.3) vanish. If $i \neq j$ and $\varepsilon_j \neq 0$, then the equality $R(e_i, e_j)e_j = h(e_j, e_j)Se_i - h(e_i, e_j)Se_j$ follows immediately from (2.7). Since the Gauss equation is satisfied, S_x is defined independently of the choice of an h-adapted basis. Moreover, since $rk\, h > 1$, the tensor field S is smooth. Namely let $x \in M$ and let e_1, e_2 be the first vectors of an h-adapted basis of T_xM. Extend e_1, e_2 to linearly independent vector fields X_1, X_2 defined in a neighbourhood U" of x. We define $E_1 = X_1/\varepsilon_1 h(X_1, X_1)$ and $\tilde{E}_2 = X_2 - h(X_2, E_1)\varepsilon_1 E_1$. Then $e_2 = \tilde{E}_2$ at x and consequently $h(\tilde{E}_2, \tilde{E}_2) \neq 0$ in a neighbourhood $U' \subset U$" of x. It is also clear that $h(E_1, \tilde{E}_2) = 0$. If we set $E_2 = \tilde{E}_2/\varepsilon_2 h(\tilde{E}_2, \tilde{E}_2)$, then the pair E_1, E_2 is h-orthogonal and $h(E_1, E_1) = \varepsilon_1$, $h(E_2, E_2) = \varepsilon_2$ on U'. We have 1-forms γ_1, γ_2 on U' given by $\gamma_i(X) = h(X, E_i)$ for $i = 1, 2$. It is obvious that $ker\gamma_i$, for $i = 1, 2$, are (n-1)-dimensional distributions on U'. Since $E_i \notin ker\gamma_i$, $E_1 \in ker\gamma_2$ and $E_2 \in ker\gamma_1$, we see that the intersection $\mathcal{W} = ker\gamma_1 \cap ker\gamma_2$ is an (n-2)-dimensional distribution on U'. Therefore on some neighbourhood $U \subset U'$ of x there is a smooth basis $E_3, \ldots, E_n$ of $\mathcal{W}$. By using the Gauss equation we get on U:

$$R(E_i, E_1)E_1 = \varepsilon_1 SE_i \text{ for } i \geq 2 \text{ and } R(E_1, E_2)E_2 = \varepsilon_2 SE_1,$$

which proves the smoothness of S.

Now we shall define a 1-form τ satisfying Codazzi equations. For an h-adapted basis $e_1, \ldots, e_2$ of T_xM we set

$$\tau(e_i) = -\varepsilon_j Q(e_i, e_j, e_j), \tag{2.8}$$

where i is arbitrary, $j \neq i$, and $\varepsilon_j \neq 0$. Similarly as above we see that τ is well defined and satisfies the Codazzi equation I. Consequently τ_x is defined independently of a basis $e_1, \ldots, e_n$ and $x \longrightarrow \tau_x$ is smooth on M. Since the Gauss equation is satisfied, we have

$$trR(X, Y) = h(Y, SX) - h(X, SY). \tag{2.9}$$

The first Codazzi equation yields

$$\sigma_{XYZ} \nabla Q(X, Y, Z, W) = 2\sigma_{XYZ} h(Y, W) d\tau(X, Z). \tag{2.10}$$

By combining this formula with assumption iii) we get

$$2\sigma_{XYZ} h(Y, W) d\tau(X, Z) = \sigma_{XYZ} h(Y, W) trR(X, Z). \tag{2.11}$$

Since $rk\, h > 2$, for arbitrary j, k we can choose i such that $i \neq j$, $i \neq k$ and $\varepsilon_i \neq 0$. If we substitute in (2.10) $X = e_j$, $Y = e_k$ and $W = Z = e_i$, then we get

$$2d\tau(e_j, e_k) = tr R(e_k, e_j), \tag{2.12}$$

which together with (2.9) gives the Ricci equation. The second Codazzi equation can be obtained in a similar way by using the second Bianchi identity for R and the Gauss equation. Now Theorem 2.1 yields the desired result.

By an easy computation one gets the following

Lemma. Let V be an n-dimensional vector space ($n > 1$) endowed with a non-degenerate symmetric bilinear form h and a tensor K of type (0,3) or (1,3). If μ is a (0,1) (resp. (1,1))-tensor on V satisfying the equation

$$K(X, Y, Z) = h(X, Z)\mu(Y) - h(Y, Z)\mu(X)$$

for every $X, Y, Z \in V$, then

$$\mu(Y) = \frac{1}{n-1} tr_h K(\cdot, Y, \cdot). \tag{2.13}$$

We now prove a theorem of Radon's type for surfaces.

Theorem 2.3. Let M be a 2-dimensional connected and simply connected manifold equipped with a torsion-free connection ∇ and a non-degenerate symmetric bilinear form h. If the following conditions are fulfilled:

ν) $tr_h(\nabla.R)(X, Y)\cdot - tr_h\nabla h(\cdot, Y, \cdot) tr_h R(X, \cdot)\cdot + tr_h \nabla h(\cdot, X, \cdot) tr_h R(Y, \cdot)\cdot$
$= tr_h Q(\cdot, X, \cdot) tr_h R(X, \cdot)\cdot - tr_h Q(\cdot, X, \cdot) tr_h R(Y, \cdot)\cdot$,

$\nu\nu$) $tr_h \nabla Q(\cdot, X, Y, \cdot) - tr_h \nabla h(\cdot, X, \cdot) tr_h Q(\cdot, Y, \cdot) + tr_h \nabla h(\cdot, Y, \cdot) tr_h Q(\cdot, X, \cdot)$
$= tr R(X, Y)$,

then there is an immersion $f : M \longrightarrow \mathbf{R}^3$ and a vector field ξ transversal to f such that ∇ and h are induced by (f, ξ).

Proof. Let $x \in M$ and e_1, e_2 be an h-orthonormal basis of $T_x M$. If we define S by

$$Se_1 = \varepsilon_2 R(e_1, e_2)e_2, \quad Se_2 = \varepsilon_1 R(e_2, e_1)e_1,$$

then S satisfies the Gauss equation. Therefore S_x is defined indepently of the choice of a basis e_1, e_2 and $x \longrightarrow S_x$ is smooth. Similarly by setting

$$\tau(e_1) = \varepsilon_2 Q(e_2, e_1, e_2), \quad \tau(e_2) = \varepsilon_1 Q(e_1, e_2, e_1)$$

we get a smooth 1-form τ on M satisfying the first Codazzi equation. The lemma applied to $K(X, Y, Z) = R(X, Y)Z$ and to $K = Q$ yields

$$SY = tr_h R(Y, \cdot)\cdot \tag{2.14}$$

and

$$\tau(Y) = tr_h Q(\cdot, Y, \cdot). \tag{2.15}$$

By the Gauss equation we obtain

$$tr_h(\nabla.R)(X,Y)\cdot - tr_h\nabla h(\cdot,Y,\cdot)SX + tr_h\nabla h(\cdot,X,\cdot)SY = \nabla S(Y,X) - \nabla S(X,Y),$$

which together with assumption ν) and formulas (2.14), (2.15) imply the second Codazzi equation. The first Codazzi equation gives

$$tr_h\nabla Q(\cdot,X,Y,\cdot) - tr_h\nabla h(\cdot,X,\cdot)\tau(Y) + tr_h\nabla h(\cdot,Y,\cdot)\tau(X) = 2d\tau(X,Y). \tag{2.16}$$

Now by combining formula (2.16), assumption $\nu\nu$) and formula (2.9), which is a consequence of the Gauss equation, we get the Ricci equation. Now Theorem 2.1 finishes the proof.

Remark. Fundamental problems for complex hypersurfaces on $\mathbf{C}^{n+1}$ will be considered in another paper.

References.

[B] Blaschke, W.: Vorlesungen über Differentialgeometrie II, Affine Differentialgeometrie, Berlin, Springer 1923.

[DNV] Dillen, F., Nomizu, K., Vrancken, L.: Conjugate connections and Radon's theorem in affine differential geometry, Monatshefte für Math. 109 (1990), 221-235.

[Gr] Griffiths, P.: On Cartan's method of Lie groups and moving frames as applied to uniqueness and existence questions in differential geometry, Duke Math. J.,41 (1974) 775-814.

[N] Nomizu, K.: Introduction to affine differential geometry, prepint.

[O] Opozda, B.: Equivalence problems in affine differential geometry, prepint.

[Sch] Schirokow, P.A., Schirokow, A.P.: Affine Differentialgeometrie, Teubner, Leipzig, 1962.

This paper is in final form and no version will appear elsewhere.

Instytut Matematyki UJ
ul. Reymonta 4
30-059 Krakow
Poland

Generalized Killing spinors with imaginary Killing function and conformal Killing fields

Hans–Bert Rademacher

1 Statement of results

We consider complete Riemannian spin manifolds (M, g) with complex spinor bundle S. S carries a hermitian product $\langle .,. \rangle$ which we assume to be complex conjugate linear in the first argument and complex linear in the second argument. The Clifford bundle $Cl(M)$ of M acts on S by Clifford multiplication which we denote by $X \cdot \psi$ for a vector field X and a spinor ψ . Clifford multiplication by a tangent vector X is skew symmetric with respect to $\langle .,. \rangle$. The Levi–Civita connection ∇ on M induces the spinor connection on S which we also denote by ∇, cf. [1]. A non–trivial spinor ψ is called *generalized Killing spinor* with *Killing function* λ if

$$\nabla_X \psi = \lambda X \cdot \psi \tag{1}$$

for all vector fields X and for a complex–valued function λ on M . In particular ψ is a *twistor spinor* , i.e. $\nabla_X \psi + \frac{1}{n} X \cdot D\psi = 0$ for all X , where D is the Dirac operator. If λ is constant then ψ is a *Killing spinor* with *Killing number* λ and (M^n, g) is an Einstein manifold of scalar curvature $r = 4n(n-1)\lambda^2$, see [8] or [5]. Hence three cases occur: If $\lambda = 0$ then ψ is parallel and M is Ricci flat. If $\lambda^2 > 0$ then λ is real, M is compact and $\lambda^2 n^2 = rn/(4n-4)$ is the smallest eigenvalue of the square D^2 of the Dirac operator D by results of T.Friedrich [9] and O.Hijazi [12]. If $\lambda^2 < 0$ then λ is an imaginary number. H.Baum classified in [2], [3] and [4] these manifolds, see Corollary 1. From results of O.Hijazi [12, cor.3.6] and A.Lichnerowicz [16, thm.1] it follows that a generalized Killing spinor is either a Killing spinor with real Killing number or λ is an imaginary function.

From now on we consider the second case, i.e. we assume $\lambda = ib$ for a not everywhere vanishing real function b.The function $f := \langle \psi, \psi \rangle$ is positive everywhere since equation (1) is a first order linear ordinary differential equation along a geodesic, cf. [16, prop.1]. The vector field V on M defined by $\langle V, X \rangle = i\langle \psi, X \cdot \psi \rangle$ for all X is a conformal non–isometric closed Killing field, cf. §3 .

T.Friedrich introduced in [10] the function $q_\psi := f^2 - ||V||^2$ which is a non–negative constant. In [17, thm.] , [18, thm.4] A.Lichnerowicz proved that locally M is the warped product of an open interval and a manifold carrying a parallel spinor if $q_\psi = 0$ and that M is globally isometric to a warped product of the real line $\mathbb{R}$ with a complete manifold carrying a parallel spinor if $q_\psi = 0$ and b has no zero [17, prop.2] , [18, prop.2] .

We obtain the above quoted results of H.Baum and A.Lichnerowicz and the global structure of the manifold M from the classification of complete Riemannian manifolds carrying a non–isometric conformal closed Killing field. This classification is given in theorem 2, related results are due to H.W.Brinkmann [7], Y.Tashiro [19], J.P. Bourguignon [6], Y.Kerbrat [13] and W.Kühnel [14] [15]. In [15] W.Kühnel studies the complete Riemannian manifolds of constant scalar curvature carrying a non–isometric conformal closed Killing field. Corollary 2 b) contains an explicit example.

Our main result is

Theorem 1 *Let M be a complete Riemannian spin manifold with a generalized Killing spinor with an imaginary Killing function, i.e.*

$$\nabla_X \psi = ib\, X \cdot \psi$$

with a not everywhere vanishing real function b .

a) If $q_\psi = 0$, then there is a positive function h on $\mathbb{R}$ and a complete $(n-1)$–dimensional Riemannian spin manifold (M_*, g_*) carrying a parallel spinor such that the warped product $\mathbb{R} \times_h M_*$ (with metric $g = du^2 + h^2(u) g_*$) is a Riemannian covering of M . Here $f = \langle \psi, \psi \rangle$ and b are functions of $u \in \mathbb{R}$ alone, $f(u,x) = f(u) = h(u), (u,x) \in \mathbb{R} \times M_*$ and $b = f'/(2f)$.

If M is a proper quotient of $\mathbb{R} \times_f M_*$ then f is periodic with period $\omega > 0$ and there is an isometry γ of M_* such that M is isometric to $\mathbb{R} \times_f M_*/\Gamma$ where the group $\Gamma \cong \mathbb{Z}$ of isometries is generated by $(u,x) \mapsto (u+\omega, \gamma(x))$.

b) If $q_\psi > 0$, then M is isometric to the n–dimensional hyperbolic space $\mathrm{H}^n(-4b^2)$ of constant sectional curvature $-4b^2$.

Hence $q_\psi = 0$ iff the conformal closed vector field V has no zero, i.e. is inessential. A conformal Killing field is inessential, if it becomes an isometric Killing field after a conformal change of the metric.

H.Baum shows in [3, thm.1] that the n–dimensional hyperbolic space carries Killing spinors with imaginary Killing number with $q_\psi = 0$ for all n and with $q_\psi > 0$ if $n \neq 3, 5$. It follows from [3, lem.4] that the warped product $\mathbb{R} \times_f M_*$ of a manifold M_* carrying a parallel spinor with an arbitrary positive function f on $\mathbb{R}$ carries a generalized Killing spinor ψ with imaginary Killing function ib with $q_\psi = 0$, where $b = f'/(2f)$.

A.Lichnerowicz describes in [17, §5] , [18, §10] the following example : Let M_* be a compact manifold carrying a parallel spinor (e.g. a K_3–surface with the Calabi–Yau metric or a flat torus with the canonical spin structure) and let $f : S^1 \longrightarrow \mathbb{R}^+$ be a positive non–constant periodic function and $b = f'/(2f)$. Then $M = S^1 \times_f M_*$ is a compact spin manifold with a generalized Killing spinor with Killing function ib. From theorem 1 it follows that up to Riemannian quotients and twisting these are all such compact manifolds.

In [11] K.Habermann gives another characterization of hyperbolic space, she shows that a complete n–dimensional Einstein spin manifold with negative scalar curvature $r = kn(n-1)$ and a non–parallel twistor spinor ψ whose length function $f = \langle \psi, \psi \rangle$ attains a minimum is the hyperbolic space $\mathrm{H}^n(k)$.

From theorem 1 we obtain

Corollary 1 (H.Baum [2], [3]) *If (M,g) is a complete Riemannian manifold carrying a Killing spinor ψ with imaginary Killing number $ib, b \in \mathbb{R} - \{0\}$ then M is isometric to*

a) $\mathbb{R} \times_{\exp(2bu)} M_*$ where M_* is a complete $(n-1)$–dimensional Riemannian manifold carrying a parallel spinor, if $q_\psi = 0$.

b) Hyperbolic space $\mathrm{H}^n(-4b^2)$, if $q_\psi > 0$.

From the formula for the scalar curvature of a warped product we obtain

Corollary 2 *Let M be a complete Riemannian manifold which carries a generalized Killing spinor with imaginary Killing function ib , $b \not\equiv 0$.*

a) There is a point of negative scalar curvature.

b) If b is non–constant, if the scalar curvature r is constant and $a := 1/2(-rn/(n-1))^{1/2}$ then M is isometric to $\mathbb{R} \times_f M_*$ where M_* carries a parallel spinor and $f(u) = (\cosh(au))^{2/n}$, hence $b(u) = a \tanh(au)/n$.

Acknowledgment: I am grateful to J.P.Bourguignon, O.Hijazi, W.Kühnel and T.Friedrich for helpful comments on the first version of this paper.

2 Conformal Killing fields

We denote by $L_V g$ the Lie derivative of the metric $g = \langle .,. \rangle$ in direction of the vector field V, i.e. $L_V g(X,Y) = \langle \nabla_X V, Y \rangle + \langle X, \nabla_Y V \rangle$. A vector field V is a *conformal Killing field* if the local flow consists of conformal transformations. This is equivalent to $L_V g = 2hg$ with a function h. By taking traces one obtains $h = \text{div} \ V/n$. V is *homothetic* if h is a constant and it is *isometric* if $L_V g = 0$. V is *closed* if the corresponding 1–form $\omega = \langle V, . \rangle$ is closed . Hence if V is a conformal closed Killing field then for every point $p \in M$ there is a neighborhood U and a function F on U such that $V = \nabla F$ on U . The Hessian $\nabla^2 F(X,Y) := \langle \nabla_X \nabla F, Y \rangle = 1/2 \ L_{\nabla F} g(X,Y)$ then satisfies

$$\nabla^2 F = \frac{\Delta F}{n} g \tag{2}$$

where Δ is the Laplacian. H.W.Brinkmann showed in [7, §3] that nearby a regular point of F the metric has a warped product structure. Y.Tashiro classifies in [19, lem.2.2] the complete Riemannian manifolds with a non–constant function F on M satisfying equation (2), i.e. ∇F is a conformal Killing field, cf. also W.Kühnel [14, thm.22] . Using this result we show

Theorem 2 *Let (M^n, g) be a complete Riemannian manifold with a non–isometric conformal closed Killing field V and let N be the number of zeros of V. Then $N \leq 2$ and:*

a) If $N = 2$, then M is conformally diffeomorphic to the standard sphere S^n.

b) If $N = 1$, then M is conformally diffeomorphic to euclidean space $\mathbb{R}^n$.

c) If $N = 0$: Then there is a complete $(n-1)$–dimensional Riemannian manifold (M_*, g_*) and a function $h : \mathbb{R} \to \mathbb{R}^+$ such that the warped product $\mathbb{R} \times_h M_*$ is a Riemannian covering of M and the lift of V is $h\frac{\partial}{\partial u}$.

If M is a proper quotient of $\mathbb{R} \times_h M_*$ then h is periodic with period $\omega > 0$ and there is an isometry γ of M_* such that $M = \mathbb{R} \times_h M_*/\Gamma$, where the group $\Gamma \cong \mathbb{Z}$ of isometries is generated by $(u,x) \mapsto (u+\omega, \gamma(x))$.

Proof . Let $\bar{M}$ be the universal Riemannian covering of M with projection $\pi : \bar{M} \to M$, denote by $\bar{V}$ the lift of V onto $\bar{M}$ and by $\bar{N}$ the number of zeros of $\bar{V}$. If G is the group of deck transformations of $\bar{M}$ such that $M = \bar{M}/G$, then $\bar{N} = \text{ord}(G)N$. Since $\bar{V}$ is closed there is a non–constant function F on $\bar{M}$ with $\bar{V} = \nabla F$ and $\nabla^2 F = (\Delta F/n)g$.

From Tashiro's classification [19, lem.2.2] resp. [14, thm.21] it follows that $\bar{N} \leq 2$ and that the following cases occur:

a) $\bar{N} = 2$, then $\bar{M}$ is conformally diffeomorphic to S^n and since $\Delta F = \text{div}\ \bar{V}$ has different signs in the critical points of F the vector field $\bar{V}$ does not project onto a proper quotient of S^n .

b) $\bar{N} = 1$, then $\bar{M}$ is conformally diffeomorphic to $\mathbb{R}^n$ and since $N = 1$ we have $M = \bar{M}$.

c) $\bar{N} = 0$, then $\bar{M}$ is isometric to $\mathbb{R} \times_{F'} \bar{M}_*$ for a complete $(n-1)$–dimensional Riemannian manifold $\bar{M}_*$ where F is a function of u alone, i.e. $F(u,x) = F(u)$ and $\bar{V} = F'\frac{\partial}{\partial u}$. Hence F' is also a function on M. $\bar{M}_*$ is a connected component of the submanifold $F^{-1}(F(p))$ for a point $p \in \bar{M}$ with $F'(p) \neq 0$. Since V is not–isometric we can assume in addition that $F''(p) = \text{div}\ \bar{V}(p)/n \neq 0$. Then $M_* := \pi(\bar{M}_*)$ is a connected component of the submanifold $F'^{-1}(F'(\pi(p))$ on M, $\mathbb{R} \times_h M_*$ is a Riemannian covering of M and the projection $\pi_1 : \mathbb{R} \times_h M_* \to M$ can be identified with the normal exponential map of the submanifold M_* in M where $h = F'$.

If M is a proper quotient then h is periodic since $\bar{V} = h\frac{\partial}{\partial u}$ is the lift of V. Let $\omega > 0$ be the greatest number such that the restriction $\pi_1 | (-\omega/2, \omega/2) \times_h M_*$ is injective, then $\gamma : M_* \to M_*$ is defined by $\gamma(x) = \pi_1(-\omega, x)$.

Remark 1 a) The cases a) and b) of theorem 2 are proved by Y.Kerbrat [13]. J.P.Bourguignon proves these cases in [6] under the additional assumption that the vector field is complete. Theorem 2 can also be found in W.Kühnel [15] .

b) If in case c) of theorem 2 M is a proper quotient then : Either the isometry γ is of finite order, then $S^1 \times_h M_*$ is a Riemannian covering and all geodesics normal to M_* are closed or otherwise no normal geodesic closes.

3 Generalized Killing spinors with imaginary Killing function

(3.1) We assume that ψ is a generalized Killing spinor with imaginary Killing function ib. From the definitions $f := \langle\psi,\psi\rangle$ and $\langle V, X\rangle := i\langle\psi, X\cdot\psi\rangle$ it follows immediately that

$$\nabla f = 2bV \quad , \quad \nabla_X V = 2bfX \quad . \tag{3}$$

Hence V is closed and since $L_V g = 4bfg$ we have that V is a non–isometric conformal Killing field. It follows from equation (3) that $q_\psi := f^2 - ||V||^2$ is constant. Let

$$Q(X) := ||X\cdot\psi - i\psi||^2 = ||X||^2 f + 2\langle V, X\rangle + f \ ,$$

then X is a minimum of Q if $X = -V/f$ and $q_\psi = fQ(-V/f) \geq 0$. Hence q_ψ is a non–negative constant and if $q_\psi = 0$ then $(-V/f)\cdot\psi = i\psi$, cf. [10].

(3.2) Let $(e_1,\ldots,e_n)$ be a local orthonormal frame, then $D\psi = \sum_{i=1}^n e_i \cdot \nabla_{e_i}\psi$ is the *Dirac operator*, $\nabla^*\nabla\psi = -\sum_{i=1}^n \nabla_{e_i}\nabla_{e_i}\psi + \nabla_{\nabla_{e_i}e_i}\psi$ is the *connection Laplacian* . We obtain

$$D\psi = -ibn\psi \ , \ D^2\psi = -b^2n^2\psi - in\nabla b\cdot\psi$$

and

$$\nabla^*\nabla\psi = -b^2n\psi - i\nabla b\cdot\psi \ .$$

Then one obtains from Lichnerowicz's formula $D^2 = \nabla^*\nabla + \frac{1}{4}r$ where r is the scalar curvature that $\nabla b \cdot \psi = i\left(b^2 n + r/(4n-4)\right)\psi$. Hence

$$|\langle \nabla b, V\rangle| = |i\langle \nabla b \cdot \psi, \psi\rangle| = \|\nabla b\| f \leq \|\nabla b\|\|V\|.$$

If b is non-constant then it follows that $q_\psi \leq 0$ i.e. $q_\psi = 0$.

Now we prove theorem 1 stated in the first section

Proof of Theorem 1.

a) If $q_\psi = 0$ then $\|V\| = f$ has no zero. By theorem 2c) M has the warped product $\mathbb{R} \times_f M_*$ of $\mathbb{R}$ with a complete $(n-1)$-dimensional manifold M_* as a Riemannian covering. Here $f(u,x) = f(u)$ is a function of $u \in \mathbb{R}$ alone and $V = f\frac{\partial}{\partial u}$ is the lift of V. Since $\nabla f = f'\frac{\partial}{\partial u} = 2bV = 2bf\frac{\partial}{\partial u}$ by equation (3) also b is a function of u alone and $b = f'/(2f)$.

Since

$$g = du^2 + f^2(u)g_* = f^2(u)(dv^2 + g_*)$$

with $\frac{dv}{du} = \frac{1}{f^2(u)}$ we compare the conformally equivalent metrics g and $\overline{g} = dv^2 + g_*$, i.e. $\overline{g}$ is the product metric on $\mathbb{R} \times M_*$. Let $f = \exp(-h)$. $\overline{g}$ induces on $\mathbb{R} \times M_*$ a spinor bundle $\overline{S}$, such that $S \longrightarrow \overline{S}, \phi \mapsto \overline{\phi} = \exp(h/2)\phi$ is an isometry. Let $\psi_1 := \exp(h/2)\psi$, then it follows from the formula [1, 3.2.4] :

$$\overline{\nabla}_X \overline{\psi_1} = \overline{\nabla_X \psi_1} - \frac{1}{2}\overline{X \cdot \nabla h \cdot \psi_1} - \frac{1}{2}X(h)\overline{\psi_1}$$

that $\overline{\psi_1}$ is a parallel spinor of $\mathbb{R} \times M_*$, cf. [17, §4]. This implies that M_* carries a parallel spinor , cf. [3, lem.4] .

b) If $q_\psi > 0$ then b is a non-zero constant by (3.2), i.e. ψ is a Killing spinor with imaginary Killing number b and (M,g) is an Einstein manifold with negative scalar curvature $r = -4n(n-1)b^2$. From equation (3) it follows that $\nabla^2 f = 4b^2 fg$, hence $\nabla f = 2bV$ is a non-homothetic conformal Killing field. If f has no critical point then by theorem 2 resp. the classification by Y.Tashiro [19, lem.2.2] we have that M is isometric to $\mathbb{R} \times_{f'} M_*$ where f is a function of $u \in \mathbb{R}$ alone. f satisfies $f'' = 4b^2 f$ and since f and f' both have no zero $f' = 2bf$. Then $q_\psi = 0$.

Hence f has a critical point, so by theorem 2 M is conformally diffeomorphic to a simply-connected space of constant sectional curvature. Since M is Einstein with $r = -4n(n-1)b^2$ it follows that M is isometric to $\mathrm{H}^n(-4b^2)$.

Remark 2 Since in theorem 2 M_* is Ricci flat it follows from the formulae for the curvature tensor of a warped product that the scalar curvature r of M is given by

$$\begin{aligned} r &= -(n-2)(n-1)\frac{f'^2}{f^2} - 2(n-1)\frac{f''}{f} && (4) \\ &= -4n(n-1)b^2 - 4(n-1)b' && (5) \end{aligned}$$

Proof of Corollary 2 .

a) Let $y := f^{n/2}$ then one obtains from equation (4) $y'' + rny/(4n-4) = 0$. If $r \geq 0$ then y has a zero since f is non-constant. This contradicts $f > 0$.

b) For constant r it follows from a) that $r < 0$. Let $a := 1/2(-rn/(n-1))^{1/2}$, i.e. $y'' - a^2 y = 0$. Since $y'/y = nb$ is non-constant y' has a zero . Let $y'(0) = 0$ then $y(u) = y(0)\cosh(au)$. By scaling g_* we can assume $y(0) = 1$.

References

[1] H.Baum: *Spin strukturen und Dirac-Operatoren über pseudo-riemannschen Mannigfaltigkeiten.* Teubner Verlag Leipzig 1981

[2] H.Baum: *Odd-dimensional Riemannian manifolds with imaginary Killing spinors.* Ann. Global Analysis Geom. **7** (1989) 141–154

[3] H.Baum: *Complete Riemannian manifolds with imaginary Killing spinors.* Ann. Global Analysis Geom. **7** (1989) 205–226

[4] H.Baum: *Variétés riemanniennes admettant des spineurs de Killing imaginaires.* C.R.Acad.Sci. Paris **309** (1989) 47–49

[5] H.Baum, T.Friedrich, R.Grunewald & I.Kath: *Twistor and Killing spinors on Riemannian manifolds.* Seminarbericht Nr. **108** Sektion Math. Humboldt–Universität Berlin 1990

[6] J.P.Bourguignon: *Transformations infinitésimales conformes fermées des variétés riemanniennes connexes complètes.* C.R.Acad.Sci. Paris **270** (1970) 1593–1596

[7] H.W.Brinkmann: *Einstein spaces which are mapped conformally on each other.* Math.Ann. **94** (1925) 119–145

[8] M.Cahen, S.Gutt, L.Lemaire & P.Spindel: *Killing spinors.* Bull. Soc. Math. Belgique **38A** (1986) 75–102

[9] T.Friedrich: *Der erste Eigenwert des Dirac Operators einer kompakten Riemannschen Mannigfaltigkeit nicht negativer Skalarkrümmung.* Math. Nachr. **97** (1980) 117–146

[10] T.Friedrich: *On the conformal relation between twistor and Killing spinors.* Suppl.Rend.Circ.Mat.Palermo (1989) 59–75

[11] K.Habermann: *The twistor equation on Riemannian manifolds.* Preprint **239** Humboldt–Universität Berlin 1989

[12] O.Hijazi: *A conformal lower bound for the smallest eigenvalues of the Dirac operator and Killing spinors.* Comm. in Math. Physics **104** (1986) 151–162

[13] Y.Kerbrat: *Existence de certains champs de vecteurs sur le variétés riemanniennes complètes.* C.R.Acad.Sci. Paris **270** (1970) 1430–1433

[14] W.Kühnel: *Conformal transformations between Einstein spaces.* In *Conformal geometry* ed. by R.S.Kulkarni and U.Pinkall. aspects of math. E 12 , Vieweg , Braunschweig 1988, 105–146

[15] W.Kühnel: *Remarks on conformal vector fields* Preprint Universität Duisburg

[16] A.Lichnerowicz: *Spin manifolds, Killing spinors and Universality of the Hijazi inequality.* Lett. in Math.Physics **13** (1987) 331–344

[17] A.Lichnerowicz: *Sur les résultats des H.Baum et Th.Friedrich concernant les spineurs de Killing à valeur propre imaginaire.* C.R.Acad.Sci.Paris **309** (1989) 41–45

[18] A.Lichnerowicz: *On the twistor spinors.* Lett. in Math.Physics **18** (1989) 333–345

[19] Y.Tashiro: *Complete Riemannian manifolds and some vector fields.* Trans. Am.Math.Soc. **117** (1965) 251–275

MATHEMATISCHES INSTITUT DER UNIVERSITÄT BONN
WEGELERSTR.10, D–W–5300 BONN 1 , FEDERAL REPUBLIC OF GERMANY

This paper is in final form and no version will appear elsewhere.

ON PROLONGATION AND INVARIANCE ALGEBRAS IN SUPERSPACE

Vladimir Rosenhaus

A b s t r a c t

The Lie procedure of finding the invariance algebra (group) of partial differential equations is generalized to the case of equations in superspace. The expressions for twice-prolonged infinitesimal operators in the superfield form are obtained. Within the frames of the approach the hidden symmetries of supersymmetric quantum mechanics are found.

1. Introduction

The present paper is devoted to the study of classical invariance groups for differential systems defined on a supermanifold. At present the theory of symmetry groups for usual differential systems with even variables, based on the works by Sophus Lie (see,e.g.[1-2]), is developed in many aspects. To mention some of them, groups of tangent transformations of infinite order (Lie Bäcklund transformation groups or generalized symmetries), a geometrical approach proceeding from E.Cartan's works (e.g.[3,4]), nonclassical symmetries [5-7], the formulation of symmetry groups in terms of jet bundles and cohomology [8], nonlocal symmetries, etc.

The aim of the present paper is to extend the logic of the theory of symmetry groups to differential systems with the Grassmann variables, whereby we are especially interested in supersymmetry equations of the second order in the superfield form, having in mind the possible physical applications. The central point of the theory of invariance groups is the prolongation theory, including the prolongation of the space, mapping, transformation, the vector field, the total derivative and the infinitesimal operator. The basis of the applications of symmetry groups is the expression for a prolonged infinitesimal operator (generator) of the group.

Let us consider a system $\mathcal{S}$ of differential equations of the k-th order

$$\mathcal{S}: \quad \omega_\nu (x^i, u^\mu_{(k)}) = 0 , \qquad \mu,\nu = 1,\dots, m, \quad i=1,\dots, n \quad , \tag{1}$$

where $u^\mu_{(k)}$ are the derivatives of u^α up to k-th order. $\mathcal{S}$ determines a submanifold $\mathcal{S}$ of the total jet space.

Let $\mathcal{H}$ be an invariance group of $\mathcal{S}$

$$\mathcal{H}: \quad x^{i'} = f^i(x,u,a) \qquad u^{\alpha'} = g^\alpha(x,u,a) \quad , \tag{2}$$

where a is the group parameter. (For simplicity we consider the one-parameter group, the result can be easily generalized). The k-th prolongation of the group $\mathcal{H}$ action will leave the submanifold $\mathcal{S}_\omega$ invariant.

Let us consider the infinitesimal operator corresponding to $\mathcal{H}$

$$\mathcal{X} = \xi^i(x,u) \frac{\partial}{\partial x^i} + \eta^\alpha(x,u) \frac{\partial}{\partial u^\alpha} \quad ,$$

$$\xi^i(x,u) = \frac{d}{da}\Big|_{a=0} f^i(x,u,a) \quad , \qquad i = 1,\dots,n \quad ,$$

$$\eta^\alpha(x,u) = \frac{d}{da}\Big|_{a=0} g^\alpha(x,u,a) \quad , \qquad \alpha = 1,\dots,m \quad .$$

The first prolongation of $\mathcal{X}$ is of the form

$$\tilde{\mathcal{X}} = \mathcal{X} + \zeta^\alpha_i \frac{\partial}{\partial p^\alpha_i} \quad , \qquad p^\alpha_i \equiv u^\alpha_{,i} \quad .$$

The standard approach [1,2] with the aid of which the expressions for the prolonged infinitesimal operator of any order can be obtained, consists in finding the component ζ^α_i (and the components of higher prolonged vector fields)

$$\zeta^\alpha_i = \frac{d}{da}\Big|_{a=o} \frac{\partial u'^\alpha}{\partial x'^i}$$

by expressing u' as a function of x' from (2) with the subsequent differentiation of the expression obtained with respect to the parameter a. However, in the superspace, to avoid the difficulties with the rigorous definition of the inverse operation, we follow (formally) a slightly different way.

2. Prolongation of the infinitesimal operator

Let us consider a partial differential equation of the second order

$$\omega\,(\,x^i,\ \theta^a\,,\ \Phi^A,\ \Phi^A_{,\mu}\,,\ \Phi^A_{,\mu\nu}\,) = 0 \quad , \tag{3}$$

where x^i $(i = 1,\dots,n)$ are usual space (even) variables, θ^a $(a = 1,\dots,l)$ are odd variables, $\left\{\theta^a, \theta^b\right\} = 0$, $\Phi^A = \Phi^A(Z^\mu)$, $A = 1,\dots,N$ are superfields, $Z^\mu = (x^i,\theta^a)$, $\mu = 1,\dots,n+1$,

$$\Phi^A_{,\mu} = \frac{\partial\Phi^A}{\partial Z^\mu} \quad , \qquad \Phi^A_{,\mu\nu} = \frac{\partial^2\Phi^A}{\partial Z^\mu\partial Z^\nu} \quad .$$

Let us consider a 1-parameter group H which is a subgroup of the invariance group of equation (3)

$$\begin{aligned} x'^i &= f^i(\,x,\ \theta,\ \Phi,\ \varepsilon\,) \quad , & i &= 1,\dots,n \ , \\ \theta'^a &= e^a(\,x,\ \theta,\ \Phi,\ \varepsilon\,) \quad , & a &= 1,\dots,l \ , \\ \Phi'^A &= g^A(\,x,\ \theta,\ \Phi,\ \varepsilon\,) \quad , \end{aligned} \tag{4}$$

where ε is the group parameter, e^a are odd and f^i, g are even functions. Let parameter ε be also even . The infinitesimal operator corresponding to the group H is of the form

$$X = \xi^i(x,\theta,\Phi)\,\frac{\partial}{\partial x^i} + \xi^{\theta^a}(x,\theta,\Phi)\,\frac{\partial}{\partial\theta^a} + \eta^A(x,\theta,\Phi)\,\frac{\partial}{\partial\Phi^A} \quad , \tag{5}$$

where

$$\xi^j = \left.\frac{\partial f^j}{\partial\varepsilon}\right|_{\varepsilon=0} \quad , \qquad \xi^{\theta^a} = \left.\frac{\partial e^a}{\partial\varepsilon}\right|_{\varepsilon=0} \quad , \qquad \eta^A = \left.\frac{\partial g^A}{\partial\varepsilon}\right|_{\varepsilon=0} \quad ,$$

and its first prolongation

$$\tilde{X} = X + \zeta^A_\mu\,\frac{\partial}{\partial p^A_\mu} \quad , \qquad p^A_\mu = \frac{\partial\Phi^A}{\partial Z^\mu} \quad . \tag{5'}$$

Let us write the following invariance condition:

$$g(x,\theta,\Phi,\varepsilon) = \Phi\,[f(x,\theta,\Phi,\varepsilon)\ ,\ e(x,\theta,\Phi,\varepsilon)] \quad . \tag{6}$$

The differentiation of (6) with respect to x gives

$$\begin{gathered} \frac{\partial g^A}{\partial x^i} + \frac{\partial g^A}{\partial\Phi^B}\,p^B_i = \left(\frac{\partial f^j}{\partial x^i} + \frac{\partial f^j}{\partial\Phi^B}\,p^B_i\right) p'^A_j + \left(\frac{\partial e^a}{\partial x^i} + \frac{\partial e^a}{\partial\Phi^B}\,p^B_i\right) p'^A_{\theta^a} \ , \\ p^A_i = \frac{\partial\Phi^A}{\partial x^i} \ , \quad p^A_{\theta^a} = \frac{\partial\Phi^A}{\partial\theta^a} \ , \quad p'^A_j = \frac{\partial\Phi'^A}{\partial x'^j} \ , \quad p'^A_{\theta^a} = \frac{\partial\Phi'^A}{\partial\theta^a} \end{gathered} \tag{7}$$

and the operation $\left.\frac{d}{d\varepsilon}\right|_{\varepsilon=0}$ applied to equation (7) leads to the following quantities:

$$\zeta^A_j = \left.\frac{\partial p'^A_j}{\partial\varepsilon}\right|_{\varepsilon=0} \quad , \qquad \xi^{\theta^a} = \left.\frac{\partial p'^A_{\theta^a}}{\partial\varepsilon}\right|_{\varepsilon=0} \quad .$$

Besides, at $\varepsilon=0$ we have,

$$f^j = x^j \quad , \quad e^a = \theta^a \quad , \quad p'^A_j = p_j \quad , \quad p'^A_{\theta^a} = p'^A_{\theta^a} \quad .$$

As a result we get

$$\frac{\partial\eta^A}{\partial x^i} + \frac{\partial\eta^A}{\partial\Phi^B} p^B_i = \zeta^A_j \, \delta_{ij} + \left(\frac{\partial\xi^j}{\partial x^i} + \frac{\partial\xi^j}{\partial\Phi^B} p^B_i\right) p^A_j + \left(\frac{\partial\xi^{\theta^a}}{\partial x^i} + \frac{\partial\xi^{\theta^a}}{\partial\Phi^B} p^B_i\right) p^A_{\theta^a} \, .$$

Thus,

$$\zeta^A_i = D_i \, \eta^A - \left(D_i \, \xi^j \right) p^A_j - \left(D_i \, \xi^{\theta^a} \right) p^A_{\theta^a} \quad , \tag{8}$$

where D_i is the total derivative operator

$$D_i = \frac{\partial}{\partial x^i} + \frac{\partial}{\partial\Phi^A} p^A_i \quad . \tag{9}$$

For $\zeta^A_{\theta^a}$ we obtain an analogous expression

$$\zeta^A_{\theta^a} = D_{\theta^a} \, \eta^A - \left(D_{\theta^a} \, \xi^j \right) p^A_j - \left(D_{\theta^a} \, \xi^{\theta^b} \right) p^A_{\theta^a} \quad . \tag{10}$$

Thus,

$$\zeta^A_\mu = D_\mu \, \eta^A - \left(D_\mu \, \xi^\nu\right) p^A_\nu \, ,$$

$$D_\mu = \frac{\partial}{\partial z^\mu} + \frac{\partial}{\partial\Phi} \, p_\mu \quad . \tag{11}$$

Formula (11) provides expressions for the components of the once-prolonged infinitesimal operator. The second prolongation is performed analogously and for a twice-prolonged vector field we get

$$\tilde{\tilde{X}} = \tilde{X} + \sigma^A_{\mu\nu} \frac{\partial}{\partial r^A_{\mu\nu}} \quad , \qquad \left(r^A_{\mu\nu} = \frac{\partial^2\Phi^A}{\partial z^\mu \partial z^\nu} \right)$$

$$\sigma^A_{\mu\nu} = \tilde{D}_\mu \left(\zeta^A_\nu \right) - \left(D_\mu \, \xi^\rho \right) r^A_{\rho\nu} \quad , \tag{12}$$

$$\tilde{D}_\mu = D_\mu + r^A_{\mu\rho} \frac{\partial}{\partial p^A_\rho} \quad .$$

The expressions for the n-th order prolongation can be obtained in the same way $\left(\tilde{X}^{(n)} \right.$ is the prolongation of $\left. X^{(\widetilde{n-1})} \right)$.

Let us note that the form of expressions (11)-(12) does not depend on whether the group parameter ε is even or odd. Formulas (11)-(12) are similar to those of the theory with even variables and enable one to find the whole point invariance algebra of any partial differential second-order equation with Grassmann variables.

3. Hidden Symmetries in Supersymmetric Quantum Mechanics

As an illustration of the application of the expressions obtained, let us consider the supersymmetric quantum mechanics [9-13] (See also [14]) with the action

$$S = \int dt d\bar{\theta} d\theta \left[\frac{1}{2} (\bar{D}\Phi)(D\Phi) - V(\Phi) \right] , \tag{13}$$

where

$$\Phi(t,\bar{\theta},\theta) = \varphi_\Phi(t) + \theta\bar{\psi}_\Phi(t) + \psi_\Phi(t)\bar{\theta} + A_\Phi(t)\theta\bar{\theta} \tag{14}$$

is a real superfield, φ_Φ and A_Φ are commuting (boson) variables and ψ_Φ and $\bar{\psi}_\Phi$ are anticommuting (fermion) variables, a bar stands for complex conjugation,

$$D = \frac{\partial}{\partial\theta} + i\bar{\theta}\frac{\partial}{\partial t} \quad , \quad \bar{D} = -\frac{\partial}{\partial\bar{\theta}} - i\theta\frac{\partial}{\partial t} \tag{15}$$

are covariant derivatives and $V(\Phi)$ is a superpotential.

The Euler-Lagrange equation

$$\frac{\delta L}{\delta\phi} = \frac{\partial L}{\partial\Phi} - D\left(\frac{\partial L}{\partial(D\Phi)}\right) - \bar{D}\left(\frac{\partial L}{\partial(\bar{D}\Phi)}\right) \tag{16}$$

has the form

$$\frac{1}{2}\left[D, \bar{D} \right]\Phi - V'(\Phi) = 0 . \tag{17}$$

Using $\left\{ D, \bar{D} \right\} = -2i\frac{\partial}{\partial t}$, we get the equation of motion

$$D\bar{D}\Phi + i\frac{\partial\Phi}{\partial t} - V'(\Phi) = 0 . \tag{18}$$

Let the twice-prolonged infinitesimal operator $\overset{\approx}{X}$ be of the form

$$\overset{\approx}{X} = \xi^{t}\frac{\partial}{\partial t} + \xi^{\theta}\frac{\partial}{\partial\theta} + \xi^{\bar\theta}\frac{\partial}{\partial\bar\theta} + \eta\frac{\partial}{\partial\Phi} + \zeta_{\mu}\frac{\partial}{\partial p_{\mu}} + \sigma_{D\bar D}\frac{\partial}{\partial r_{D\bar D}}\ , \tag{19}$$

where

$$z^{\mu} = \left(t,\theta,\bar\theta\right)\ ,\quad r_{D\bar D} = D\bar D\Phi\ ,\quad \sigma_{D\bar D} = \frac{\partial}{\partial\varepsilon}\left(D'\bar D'\Phi'\right)\Big|_{\varepsilon=0}\ . \tag{20}$$

Using equations (15),(14) and (11)-(12), we get

$$\begin{aligned}
\sigma_{D\bar D} &= \tilde{D}_{D}\left(\zeta_{\bar D}\right) - \left(D_{D}\xi^{\mu}\right)\left[\frac{\partial}{\partial z^{\mu}}\left(\bar D\Phi\right)\right] + i\xi^{\bar\theta}\left(\bar D p_{t}\right)\ ,\\
\zeta_{\bar D} &= D_{\bar D}\eta - \left(D_{\bar D}\xi^{\mu}\right)p_{\mu} - i\xi^{\theta}p_{t}\ ,\\
D_{\bar D} &= \bar D + \left(\bar D\Phi\right)\frac{\partial}{\partial\Phi}\ ,\\
\tilde{D}_{D} &= D_{D} + \left(Dp_{\mu}\right)\frac{\partial}{\partial p_{\mu}} = D + (D\Phi)\frac{\partial}{\partial\Phi} + \left(Dp_{\mu}\right)\frac{\partial}{\partial p_{\mu}}\ .
\end{aligned} \tag{21}$$

Thus, for the equation of supersymmetric quantum mechanics we get

$$\begin{aligned}
\omega:&\quad r_{D\bar D} + ip_{t} - V'(\Phi) = 0\ ,\\
\overset{\approx}{X}\omega\Big|_{\omega=0}:&\quad \sigma_{D\bar D} + i\zeta_{t} - \eta V''(\Phi) = 0\ .
\end{aligned} \tag{22}$$

Allowing for the independence of all (remaining) derivatives p_{μ} and $r_{\mu\nu}$ in the extended space$(t,\theta,\bar\theta,\Phi,p,r)$, we obtain from (22)

$$\begin{aligned}
\xi^{t} &= \varphi + \theta\bar\psi + \psi\bar\theta\ ,\\
\xi^{\theta} &= i\psi + R\theta - \theta\bar\theta\psi_{,t}\ ,\\
\xi^{\bar\theta} &= -i\bar\psi + (\varphi_{,t}-R)\bar\theta - \theta\bar\theta\bar\psi_{,t}\ ,
\end{aligned} \tag{23}$$

where the functions $\varphi(t)$ and $R(t)$ are commuting and $\bar\psi(t)$ and $\psi(t)$ are anticommuting, $\varphi_{,t} = \frac{\partial\varphi}{\partial t}$, and

$$\eta = A(t,\theta,\bar\theta)\Phi + B(t,\theta,\bar\theta)\ , \tag{24}$$

where A and B satisfy the equation

$$D\bar D B + iB_{,t} + (D\bar D A + iA_{,t})\Phi - (A\Phi+B)V''(\Phi) + V'(\Phi)(A - D\xi^{\theta} + \bar D\xi^{\bar\theta}) = 0. \tag{25}$$

1) For the superpotential $V(\Phi)$ of an arbitrary form we have (A = B = 0)

$$\xi^t = \varphi_0 + \theta\bar{\psi}_0 + \psi_0\bar{\theta} \quad , \qquad \eta = 0 \quad ,$$
$$\xi^\theta = i\psi_0 + R_0\theta \quad , \qquad \xi^{\bar{\theta}} = - i\bar{\psi}_0 - R_0\bar{\theta} \quad . \tag{26}$$

Thus the invariance algebra L_s in the case of an arbitrary superpotential $V(\Phi)$ is generated by

$$X_1 \equiv P = \frac{\partial}{\partial t} \quad ,$$
$$X_2 \equiv D_2 = \theta\frac{\partial}{\partial\theta} - \bar{\theta}\frac{\partial}{\partial\bar{\theta}} \quad ,$$
$$X_3 \equiv Q_1 = \frac{\partial}{\partial\theta} - i\bar{\theta}\frac{\partial}{\partial t} \quad , \tag{27}$$
$$X_4 \equiv Q_2 = \frac{\partial}{\partial\bar{\theta}} - i\theta\frac{\partial}{\partial t} \quad .$$

$$\left[X_2, X_3 \right] = - X_3 \quad , \quad \left[X_2, X_4 \right] = X_4 \quad , \quad \left\{ X_3, X_4 \right\} = -2iX_1, \tag{28}$$

X_3, X_4 are known (translational) supersymmetry generators.

The finite transformation $T_3(\varepsilon)\ T_4(\bar{\varepsilon})$ is

$$\theta' = \theta + \varepsilon \quad , \quad \bar{\theta}' = \bar{\theta} + \bar{\varepsilon} \quad , \quad t' = t - i(\varepsilon\bar{\theta} + \bar{\varepsilon}\theta) \quad . \tag{29}$$

For the superpotential $V(\Phi)$ of some special forms the invariance algebra turns out to be larger than L_s (including L_s as the subalgebra).

2) $\quad V(\Phi) = ue^{\alpha\Phi} + c \quad , \qquad \alpha \neq 0 \; , \; u, \alpha, c = \text{const.}$

The invariance algebra is generated by the 5 vector fields $X_1, \ldots, X_5$, where

$$X_5 = \frac{\alpha}{2}\left(\theta\frac{\partial}{\partial\theta} + \bar{\theta}\frac{\partial}{\partial\bar{\theta}} + 2t\frac{\partial}{\partial t} \right) - \frac{\partial}{\partial\Phi} \tag{30}$$

3) $\quad V(\Phi) = u(\Phi+\alpha)^{\beta+1} + c \quad , \qquad \beta \neq 0 \; , \; \mp 1$

The generators of invariance algebra are $X_1, \ldots, X_5$ again, where

$$X_5 = \frac{1-\beta}{2}\left(\theta\frac{\partial}{\partial\theta} + \bar{\theta}\frac{\partial}{\partial\bar{\theta}} + 2t\frac{\partial}{\partial t} \right) + (\Phi + \alpha)\frac{\partial}{\partial\Phi} \quad . \tag{31}$$

4) $V(\Phi) = u\cdot\ln(\Phi + b) + \alpha\left(\frac{\Phi^2}{2} + b\Phi\right) + c \quad , \quad \alpha \neq 0 \; , \quad u, b, \alpha, c = \text{const}.$

The corresponding 8-dimensional invariance algebra is spanned by $X_1, \dots X_8$, where

$$X_5 = e^{2i\alpha t}\left\{2\alpha\bar\theta(\Phi + b)\frac{\partial}{\partial\Phi} + \left(1 - 2\alpha\theta\bar\theta\right)\frac{\partial}{\partial\theta} - i\bar\theta\frac{\partial}{\partial t}\right\} ,$$

$$X_6 = e^{-2i\alpha t}\left\{2\alpha\theta(\Phi + b)\frac{\partial}{\partial\Phi} - \left(1 - 2\alpha\theta\bar\theta\right)\frac{\partial}{\partial\bar\theta} + i\theta\frac{\partial}{\partial t}\right\} ,$$

$$X_7 = e^{2i\alpha t}\left\{\alpha\left(1 + 2\alpha\theta\bar\theta\right)(\Phi + b)\frac{\partial}{\partial\Phi} + 2\alpha\theta\frac{\partial}{\partial\theta} - i\frac{\partial}{\partial t}\right\} , \tag{32}$$

$$X_8 = e^{-2i\alpha t}\left\{\alpha\left(1 + 2\alpha\theta\bar\theta\right)(\Phi + b)\frac{\partial}{\partial\Phi} + 2\alpha\bar\theta\frac{\partial}{\partial\bar\theta} + i\frac{\partial}{\partial t}\right\} .$$

Introducing

$$\begin{aligned}
Q_3 &= \tfrac{1}{2}\left(iX_1/\alpha - X_2\right) , & V_+ &= X_3 , \\
Q_+ &= X_8/2\alpha , & V_- &= -X_5 , \\
Q_- &= X_7/2\alpha , & W_+ &= X_6/4\alpha , \\
B &= -X_2/2 , & W_- &= X_4/4\alpha
\end{aligned} \tag{33}$$

we can see that the algebra of invariance in this case is $Osp(2,2) \simeq spl(2,1)$ (e.g.[15]).

5) $V(\Phi) = u\cdot\ln(\Phi + b) + c$

In this case the equation of supersymmetric quantum mechanics possesses a superconformal group in (1,2) space ($\simeq Osp(2,2)$) ([11]) with the generators

$$X_5 \equiv D = \tfrac{1}{2}\left(\theta\frac{\partial}{\partial\theta} + \bar\theta\frac{\partial}{\partial\bar\theta} + 2t\frac{\partial}{\partial t} + \Phi\frac{\partial}{\partial\Phi}\right) ,$$

$$X_6 \equiv K = t\left(\theta\frac{\partial}{\partial\theta} + \bar\theta\frac{\partial}{\partial\bar\theta} + t\frac{\partial}{\partial t} + \Phi\frac{\partial}{\partial\Phi}\right) ,$$

$$X_7 \equiv S_1 = i\left(t + i\theta\bar\theta\right)\frac{\partial}{\partial\theta} + \bar\theta\left(t\frac{\partial}{\partial t} + \Phi\frac{\partial}{\partial\Phi}\right) , \tag{34}$$

$$X_8 \equiv S_2 = i\left(t - i\theta\bar\theta\right)\frac{\partial}{\partial\bar\theta} + \theta\left(t\frac{\partial}{\partial t} + \Phi\frac{\partial}{\partial\Phi}\right) ,$$